Spicy and Aromatic Plants

Spicy and Aromatic Plants

Editors

Romina Alina Marc
Crina Carmen Mureșan
Andruța Elena Mureșan

MDPI • Basel • Beijing • Wuhan • Barcelona • Belgrade • Manchester • Tokyo • Cluj • Tianjin

Editors
Romina Alina Marc
University of Agricultural Science and Veterinary Medicine Cluj-Napoca
Cluj-Napoca, Romania

Crina Carmen Mureșan
University of Agricultural Science and Veterinary Medicine Cluj-Napoca
Cluj-Napoca, Romania

Andruța Elena Mureșan
University of Agricultural Science and Veterinary Medicine Cluj-Napoca
Cluj-Napoca, Romania

Editorial Office
MDPI
St. Alban-Anlage 66
4052 Basel, Switzerland

This is a reprint of articles from the Special Issue published online in the open access journal *Plants* (ISSN 2223-7747) (available at: https://www.mdpi.com/journal/plants/special_issues/Spicy_Aromatic).

For citation purposes, cite each article independently as indicated on the article page online and as indicated below:

LastName, A.A.; LastName, B.B.; LastName, C.C. Article Title. *Journal Name* **Year**, *Volume Number*, Page Range.

ISBN 978-3-0365-6982-6 (Hbk)
ISBN 978-3-0365-6983-3 (PDF)

Contents

Preface to "Spicy and Aromatic Plants"

The purpose of this book entitled "Spicy and Aromatic Plants" is to expand our knowledge of edible plants, which have been used for thousands of years by all the peoples of the world, in every household. They give special flavors to culinary preparations and have various healing properties. This special issue contains an editorial and six scientific articles (five original research articles and one original review), which contribute to the knowledge of edible plants used in fields such as the food, pharmaceutical, cosmetic and agricultural industries and are addressed to scientists and the general public.

We would like to thank you for your support during the editing of Ms. Mera Qi and the Author Service Manager. I want to thank the authors who contributed to this book. I would like to thank the technical publisher who prepared this book for publication by Plants MDPI. Also, we want to thank our husbands and children for their understanding and colleagues from the University of Agricultural Sciences and Veterinary Medicine Cluj-Napoca for their support.

Romina Alina Marc, Crina Carmen Mureșan, and Andruța Elena Mureșan
Editors

plants

MDPI

Editorial

Spicy and Aromatic Plants

Romina A. Marc, Crina C. Mureșan * and Andruța E. Mureșan *

Food Engineering Department, Faculty of Food Science and Technology, University of Agricultural Science and Veterinary Medicine Cluj-Napoca, 3-5 Calea Mănăștur Street, 400372 Cluj-Napoca, Romania
* Correspondence: crina.muresan@usamvcluj.ro (C.C.M.); andruta.muresan@usamvcluj.ro (A.E.M.)

1. Introduction

The purpose of this Special Issue entitled "Spicy and Aromatic Plants" is to expand our knowledge about edible plants, which have been used for thousands of years, by all the peoples of the world, in every household. They give special flavors to culinary preparations and have various curative properties. This Special Issue contains six scientific articles (five original research articles and one original review), which contribute the knowledge of edible plants used in fields, such as the food, pharmaceutical, cosmetic, and agricultural industries. This editorial aims to summarize the valuable work that is published in this Special Issue, and to increase the visibility and citations of these studies.

2. Overview of the Special Issue

Puscaș et al. (2022) [1] published a research paper entitled "Cytotoxicity Evaluation and Antioxidant Activity of a Novel Drink Based on Roasted Avocado Seed Powder". The paper describes the creation and utilization of avocado seed in a hot drink, similar to already-existing coffee alternatives, obtained by infusing roasted and ground avocado seeds. Different time and temperature protocols for roasting avocado seeds were evaluated, along with drying, as were changes in flavor and color. It was proposed that the powder of roasted avocado seeds be valorized in a hot drink (at 180 °C for 25 min) by making an infusion of 7% powder and hot water. Raw and conditioned avocado seeds, as well as the resulting drinks, were analyzed. Seeds have large amounts of carbohydrates, including dietary fiber. The percentage of proteins determined was 4–5%, with the difference being dependent on the process applied: drying or roasting. Flax in the raw seed was determined to have the highest content of polyphenols (772.90 mg GAE/100 g). The amount of polyphenols in the drink was much lower (17.55 mg GAE/100 g). The antioxidant capacity of the drink was high. The acidity and total carotenoid compound increased significantly during conditioning. The antioxidant capacity of the drink was high (90.27 RSA%), which could be due to the high content of total carotenoid compounds detected in the roasted seed (6534.48 μg/100 g) or flavonoids. The 7% roasted avocado seed powder drink was shown to have high antiproliferative activity in Hs27 and DLD-1 cell lines. These values are higher than the previously reported values for coffee and coffee substitutes in other studies.

Saqib et al. (2022) [2] published a research paper entitled "Antidiarrheal and Cardio-Depressant Effects of *Himalaiella heteromalla* (D.Don) Raab-Straube: In Vitro, In Vivo, and In Silico Studies". *Himalaiella heteromalla* (D.Don) Raab-Straube is commonly known as Batula. Its use dates back a long time as an adjuvant in the treatment of various diseases. In this work, the crude extract of *H. heteromalla* and its fractions were investigated for their gastrointestinal, bronchodilator, cardiovascular, and anti-inflammatory activities. *H. heteromalla* crude extract (Hh.Cr) dose-dependently relaxed spontaneous contractions and K^+ (80 mM)-induced contraction in the jejunum tissue. The relaxation of K^+ (80 mM) indicates the presence of a Ca^{++} channel blocking (CCB) effect, which was further confirmed by constructing calcium response curves (CRCs) as they caused a rightward parallel shift in CRCs in a manner comparable to verapamil; so, the spasmolytic effect of Hh.Cr was due to

Citation: Marc, R.A.; Mureșan, C.C.; Mureșan, A.E. Spicy and Aromatic Plants. *Plants* **2023**, *12*, 848. https://doi.org/10.3390/plants12040848

Received: 22 October 2022
Accepted: 9 February 2023
Published: 14 February 2023

its CCB activity. The application of Hh.Cr in CCh (1 μM)- and K^+ (80 mM)-induced contraction in tracheal preparation resulted in complete relaxation, showing its bronchodilator effect, which was mediated through Ca^{++} channels and cholinergic antagonist activity. The application of Hh.Cr in aortic preparations resulted in vasorelaxant activity through angiotensin and α-adrenergic receptor blockage. It also had a cardio-suppressant effect with negative chronotropic and inotropic responses in paired atrium preparation. Similar effects were observed in in vivo models, i.e., decreased propulsive movement, wet feces, and the inhibition of edema formation. *Himalaiella heteromalla* had a more spasmolytic effect in the ethyl acetate fraction and caused complete relaxation of the isolated jejunum, trachea, aorta, and paired atria; this was supported by in silico studies.

Ciocarlan et al. (2021) [3] published a research paper entitled "Chemical Composition and Assessment of Antimicrobial Activity of Lavender Essential Oil and Some By-Products". The purpose of the present study was to analyze the chemical composition of lavender (*Lavanda angustifolia* L.) essential oil and some by-products derived from its production (residual water and residual herbs), as well as to assess their "in vitro" antimicrobial activity. Lavender samples from seven industrial producers from Moldova were analyzed and 41 essential oil compounds were identified. The method used was gas chromatography–mass spectrometry. Significant antimicrobial activity was identified for *Bacillus subtilis, Pseudomonas fluorescens*, *Xanthomonas campestris*, *Erwinia carotovora*, *Erwinia amylovora*, and *Candida utilis*. Antimicrobial activity of lavender plant material, wastewater, and ethanol extracts from solid waste residue was observed for *Aspergillus niger*, *Alternaria alternata*, *Penicillium chrysogenum*, *Bacillus* sp., and *Pseudomonas aeruginosa*. Antimicrobial activity differed depending on the concentrations.

Mushtaq et al. (2021) [4] published a research paper entitled "Biomolecular Evaluation of *Lavandula stoechas* L. for Nootropic Activity". The aim of this study was to identify the biomolecules of lavender that are responsible for improving memory. An aqueous extract of *L. stoechas* was made, which was first purified, and then, analyzed in vitro for anticholinesterase activity. An active fraction of *L. stoechas* (AfL.s) was subjected to biomolecule analysis using the GS-MS method. Two main compounds were identified: α-tocopherol and phenethylamine. These compounds were administered to mice at different doses for two days. The mice were sacrificed and their brains were analyzed for a biochemical assay. α-Tocopherol reduces free radical oxidative stress in the brains of mice, while phenethylamine increases the levels of acetylcholine in the hippocampus. Thus, it was concluded that α-tocopherol and phenethylamine (a primary amine) present in *L. stoechas* improved the memories of the studied animals and can be used as a memory enhancer.

Sriwichai et al. (2021) [5] published a research paper entitled "Aromatic Profile Variation of Essential Oil from Dried Makwhaen Fruit and Related Species". This work was carried out to evaluate the relationships between the genotype, phenotype, and chemical profiles of essential oil obtained from Zanthoxylum spp. A morphological comparison was carried out for three specimens of Makhwaen (MK) with other Zanthoxylum spices: Huajiao (HJ) and Makwoung (MKO). The extracted essential oils were analyzed from a chemical and physical point of view. MKO and MK showed similar volatile profiles of fruity, woody, and citrus aromas, while HJ was distinctive with a citrus-floral aroma.

Marc et al. (2022) [6] published a research paper entitled "Spicy and Aromatic Plants for Meat and Meat Analogues Applications". This review paper aimed to present updated information on the antioxidant and antimicrobial properties of the most common herbs and spices (parsley, dill, basil, oregano, sage, coriander, rosemary, marjoram, tarragon, bay, thyme, and mint) used in the meat and meat analogue industries.

3. Conclusions

The growing interest of consumers, processors, and producers in spicy and aromatic plants makes this Special Issue highly interesting. These plants are appealing due to their natural aromas, antioxidants, and antimicrobial substances, as well as their demonstrated bioactive compounds. The use of these plants is recommended due to their ability to improve flavor, and for their health benefits to consumers. We are pleased that our published papers have different numbers of citations, which demonstrates their quality and the interest of readers and scientists.

Author Contributions: All authors: R.A.M., C.C.M. and A.E.M. contributed equally to this article. All authors have read and agreed to the published version of the manuscript.

Funding: This research received no external funding.

Conflicts of Interest: The authors declare no conflict of interest.

References

1. Pușcas, A.; Tanislav, A.E.; Marc, R.A.; Muresan, V.; Muresan, A.E.; Pall, E.; Cerbu, C. Cytotoxicity Evaluation and Antioxidant Activity of a Novel Drink Based on Roasted Avocado Seed Powder. *Plants* **2022**, *11*, 1083. [CrossRef] [PubMed]
2. Saqib, F.; Usman, F.; Malik, S.; Bano, N.; Ur-Rahman, N.; Riaz, M.; Marc (Vlaic), R.A.; Muresan, C.C. Antidiarrheal and Cardio-Depressant Effects of *Himalaiella heteromalla* (D.Don) Raab-Straube: In Vitro, In Vivo, and In Silico Studies. *Plants* **2022**, *11*, 78. [CrossRef]
3. Ciocarlan, A.; Lupascu, L.; Aricu, A.; Dragalin, I.; Popescu, V.; Geana, E.-I.; Ionete, R.E.; Vornicu, N.; Duliu, O.G.; Hristozova, G.; et al. Chemical Composition and Assessment of Antimicrobial Activity of Lavender Essential Oil and Some By-Products. *Plants* **2021**, *10*, 1829. [CrossRef]
4. Mushtaq, A.; Anwar, R.; Gohar, U.F.; Ahmad, M.; Marc (Vlaic), R.A.; Mureşan, C.C.; Irimie, M.; Bobescu, E. Biomolecular Evaluation of *Lavandula stoechas* L. *for Nootropic Activity. Plants* **2021**, *10*, 1259. [CrossRef]
5. Sriwichai, T.; Wisetkomolmat, J.; Pusadee, T.; Sringarm, K.; Duangmal, K.; Prasad, S.K.; Chuttong, B.; Sommano, S.R. Aromatic Profile Variation of Essential Oil from Dried Makwhaen Fruit and Related Species. *Plants* **2021**, *10*, 803. [CrossRef] [PubMed]
6. Marc, R.A.; Mureșan, V.; Mureșan, A.E.; Mureșan, C.C.; Tanislav, A.E.; Puscas, A.; Martis, G.S.; Ungur, R.A. Spicy and Aromatic Plants for Meat and Meat Analogues Applications. *Plants* **2022**, *11*, 960. [CrossRef] [PubMed]

plants

Review

Spicy and Aromatic Plants for Meat and Meat Analogues Applications

Romina Alina Marc (Vlaic) [1], Vlad Mureșan [1,*], Andruţa E. Mureșan [1,*], Crina Carmen Mureșan [1], Anda E. Tanislav [1], Andreea Pușcaș [1], Georgiana Smaranda Marţiș (Petruţ) [1] and Rodica Ana Ungur [2]

1 Food Engineering Department, Faculty of Food Science and Technology, University of Agricultural Science and Veterinary Medicine Cluj-Napoca, 3-5 Calea Mănăştur Street, 400372 Cluj-Napoca, Romania; romina.vlaic@usamvcluj.ro (R.A.M.); crina.muresan@usamvcluj.ro (C.C.M.); anda.tanislav@usamvcluj.ro (A.E.T.); andreea.puscas@usamvcluj.ro (A.P.); georgiana.petrut@usamvcluj.ro (G.S.M.)

2 Department of Rehabilitation Iuliu-Haţieganu, Faculty of General Medicine, University of Medicine and Pharmacy, 8 Victor Babes Street, 400012 Cluj-Napoca, Romania; rodica.ana.ungur@gmail.com

* Correspondence: vlad.muresan@usamvcluj.ro (V.M.); andruta.muresan@usamvcluj.ro (A.E.M.)

Abstract: Aromatic and spicy plants are an important factor that contributes not only to improving the taste of meat, meat products, and meat analogues, but also to increasing the nutritional value of the products to which they are added. The aim of this paper is to present the latest information on the bioactive antioxidant and antimicrobial properties of the most commonly used herbs and spices (parsley, dill, basil, oregano, sage, coriander, rosemary, marjoram, tarragon, bay, thyme, and mint) used in the meat and meat analogues industry, or proposed to be used for meat analogues.

Keywords: herbs; essential oils; aroma compounds; antioxidant activity; antibacterial activity; bioactive compounds

Citation: Marc, R.A.; Mureșan, V.; Mureșan, A.E.; Mureșan, C.C.; Tanislav, A.E.; Pușcaș, A.; Marţiș, G.S.; Ungur, R.A. Spicy and Aromatic Plants for Meat and Meat Analogues Applications. *Plants* **2022**, *11*, 960. https://doi.org/10.3390/plants11070960

Academic Editor: Manuel Viuda-Martos

Received: 13 March 2022
Accepted: 30 March 2022
Published: 1 April 2022

1. Introduction

Spicy and aromatic plants have been used in human consumption for thousands of years (since around 5000 BC). Initially, they played an important role in primary care, being used as therapeutic agents in the treatment of various diseases; however, wider applications are reported today [1]. Over time, spicy and aromatic plants began to be used around the world in various foods to flavor them, but also for preservative purposes. These plants are considered an untapped reservoir of valuable substances, also called phytochemicals, phytogenic, phytobiotics, botanicals or spices, although they are not established as essential ingredients [1–4].

Meat is known to be an important source of protein, essential amino acids, vitamins, and minerals. However, most of the meat worldwide is processed. After processing, the meat becomes more perishable and sensitive to oxidation. To improve these characteristics, as well as the aroma, aromatic and spicy plants with aromatizing roles and natural antioxidants are used [5,6]. Synthetic chemicals are also used, but consumers prefer natural antioxidants due to the possible long-term toxic effects of synthetic substances [7].

The growing population around the world has led to the need to increase the number of protein-containing products. Meat and meat products are the most common sources of high protein, but these sources are no longer able to meet all the needs of consumers: an increasing amount is needed, and for a part of the population, these products are not recommended for certain diseases. Along with this need, the interest in meat analogues has risen considerably [8]. The demand for these vegetable meat alternatives is growing, because they have benefits for consumers, but also for the planet, and are recognized as sustainable protein sources. A vegetable-based diet has been shown to reduce the risk of cardiovascular disease, diabetes, high blood pressure, and mortality [9].

To make these food products tasty, and to have a pleasant appearance, whether referring to meat products or meat analogues, we use herbs and spices. They are used not only for their flavor, but also for the benefits they bring to finished products and the benefits they bring to consumers. In the meat and meat analogues industries, the most commonly used aromatic and spicy plants are parsley, dill, basil, oregano, sage, coriander, rosemary, marjoram, tarragon, bay, thyme, and mint [2,10–28]. They contain chemicals such as polyphenols, flvonoids, quinics, polypeptides, and alkaloids, or their oxygen-substituted derivatives. Some of these substances can act synergistically and improve bioactivity. Additionally, these bioactive compounds have therapeutic value, such as antioxidant and antiseptic activity [1]. Thus, the active components of these plants might have the ability to reduce the risk of cancer, cardiovascular disease, respiratory disease, and stomach or inflammatory disorders, and reduce oxidative stress. They also contain antimicrobial compounds, which delay microbial growth in food [1,4]. This review aims to gather recent information on spicy and aromatic plants used to prepare meat and meat alternatives.

2. Spicy, Aromatic Plants and Their Applications in Meat and Meat Analogues

The aromatic and spicy plants regularly used in meat preparations and meat analogues are parsley, dill, basil, oregano, sage, coriander, rosemary, marjoram, tarragon, bay, thyme, and mint. Different parts of plants are used, such as stems, seeds, or leaves, in different forms, including extract powder, essential oils, ground leaves, herbal dust, water extract, or powder extract. They are used in different amounts depending on the type of product to which they are added, and have flavoring, coloring, antimicrobial, antioxidant effects, as shown in Table 1.

Table 1. The effect of different parts of spicy and aromatic plants in meat products.

Plant Name	Plant of Origin	Part of the Plant	Form Used	Quantity Used	Product Added	Effect	Reference
Parsley	*Petroselinum crispum* Hoffm.	Parsley stems	Extract powder	4.29 g/kg	Mortadella-type sausage	Inhibition of *L. Monocytogenes* and microbial spoilage during storage time	[10]
Dill	*Anethum graveolens* L.	Seeds	A water-soluble polysaccharide named AGP1 was isolated from seeds of *Anethum graveolens*	0.3% (*g*/*g*)	Turkey sausages	Preservative AGP1 replaced ascorbic acid, reduced lipid peroxidation, preserved pH and color and improved bacterial stability during cold storage at 4 °C for 12 days	[11]
Basil	*Ocimum basilicum* L.	Leaves	Essential oils	9.0 µl/ml	Fermented sausages	Reduction in mold growth; antifungal protection	[12]
				0.062–0.25%	Beef burger	Antioxidant and antibacterial activity	[13]
Oregan	*Origanum vulgare* L.	Dried leaves from Chile leaves	Essential oils	0.230–0.690 mg/ml	Sausage	Bacteriostatic effect; antimicrobial activity	[14]
				6.25–100 µl ml^{-1}	Poultry meat products	Bio-preservative; antimicrobial activity against *Staphylococcus aureus*	[15]
Sage	*Salvia officinalis* L.	Leaves	Ground sage leaves	0.05–0.15%	Chinese style sausage	Improve the oxidative stability of Chinese-style sausage antioxidant activity	[16]
			Essential oils	0.2 și 0.5%	Chicken fat	Effective in respect of hydrolytic rancidity	[17]

Table 1. *Cont.*

Plant Name	Plant of Origin	Part of the Plant	Form Used	Quantity Used	Product Added	Effect	Reference
		Sage tea processing by product	Essential oil	0.05 µl/g–0.1 µl/g	Fresh pork sausages	Significant antioxidative and antimicrobial activities	[18]
			Herbal dust	0.05 µl/g–0.1 µl/g	Fresh pork sausages	Significant antioxidative and antimicrobial activities	[18]
Coriander	*Coriandrum sativum* L.	Leaves	Essential oils	0.01%	Italian salami	Reduction in lipid oxidation by increasing the shelf life of the product	[19]
				0.02%	Stored ground beef	Inhibiting the development of unwanted sensory changes and the growth of Enterobacteriaceae	[20]
				0.075–0.150 µl/g	Cooked pork sausages	Improved oxidative stability	[21]
Rosemary	*Rosmarinus officinalis* L.	Leaves	Water extract	0.4% rosemary spice and 0.6% nitrite pickling salts	Beef sausages	Microbial inhibition;Rosemary spice can substitute nitrite pickling salt	[22]
			Extract in powder	0.025–0.05%	Fermented goat meat sausage	Oxidative stability (antioxidant activity)	[23]
			Essential oils	0.2 și 0.5%	Chicken fat	Antioxidant activity; effective in respect of hydrolytic rancidity	[17]
Marjoram	*Origanum majorana* L.	Leaves	Essential oils	6.25-100 µl ml^{-1}	Poultry meat products	Bio-preservative; antimicrobial activity against *Staphylococcus aureus*	[15]
Tarragon	*Artemisia dracunculus* L.	Leaves	Essential oils	0.062–0.25%	Beef burger	Natural preservative, flavor enhancer in meat; antioxidant activity and antibacterial effects; anti-*staphylococcus aureus* activity	[13]
				0.1% (*v*/*w*)	Frankfurter type sausages	Improving sensory properties	[24]
Bay	*Laurus nobilis* L.	Leaves	Essential oils	10% (*v*/*v*)	Fresh lamb meat	Natural preservative; antibacterial activity	[25]
				0.05 g–0.1 g/100 g	Fresh Tuscan sausage	Antibacterial activity; improve the safety and shelf life	[26]
Thyme	*Thymbra capitata* (L.) Cav	Leaves	Essential oils	0.01–3% [*v*/*w*]	Minced beef meat	Antibacterial (from *Escherichia coli*, *Salmonella typhimurium*, *Staphylococcus aureus*, *Pseudomonas aeruginosa*)	[27]
Mint	*Mentha piperita* L.	Leaves	Essential oils	20, 40 and 60 ppm	Cooked sausage	Nitrite partial replacement with *Mentha piperita* essential oil proved oxidative, microbial, and sensory properties	[28]

3. Bioactive Compounds and Antibacterial and Antioxidant Activity of Spicy and Aromatic Plants in Meat

The consumption of fresh and processed meat has an indisputable value for diet as a source of proteins and micronutrients, but in 2016, an International Agency for Research on Cancer (IARC) working group classified processed meat as "carcinogenic to humans" and red meat as "probably carcinogenic to humans for colorectal cancer". On the other hand, fresh and processed meat is an key part of the Western diet, associated with chronic metabolic inflammation and an important group of chronic diseases: obesity, hyperlipidemia, diabetes, gout, high blood pressure and degenerative neurological diseases, including dementia. In this context, it is very important to associate meat and its derivatives with compounds that can counteract their possible negative effects.

Spices can be important factors that contribute not only to improvements in meat and meat product savor, but also to increasing the nutritive value of meat and derivatives, and counteracting their metabolic disadvantages. It has been demonstrated that plants with green leaves utilized as spices for meat and derivatives have important antioxidant and anti-inflammatory effects, antidiabetic, antimicrobial and antimutagenic actions, associated with their chemical composition, i.e., rich in polyphenols, carotenoids, and terpenoids. The use of synthetic antioxidants is restricted due to their carcinogenicity; therefore, natural antioxidants derived from plants, including aromatic plants, are recommended for use in the food industry. The prevalence of digestive disorders in Western populations is currently increasing. Culinary spices used in meat preparations can stimulate digestive processes by increasing bile and digestive enzyme production, and modulating the structure and function of gut microbiota. On the other hand, the antioxidant, anti-inflammatory, antimicrobial, antifungal, and antimutagenic effects of spices protect the digestive system from cancer, gastritis and ulcers, periodontitis, and colitis. Recently, many spices used in meat processing have been proposed as alternative treatments for SARS-CoV-2-infected patients due to their anti-inflammatory properties that can be potentially efficient to combat a cytokine storm [29].

Due to their ability to prevent and slow down the rate of lipid oxidation in food systems, aromatic herbs and their derivatives are potential natural alternative sources for antioxidants and synthetic preservatives. They can be added directly to the product (in fresh or dried states), its constituents can be extracted and added in the form of essential oil (the addition is limited by the intense aroma) or extracts, and in the combination of different plant extracts. The last method can lead to superior antioxidant activity due to the synergistic action of the compounds. Natural antioxidants have the effect of reducing the formation of cytotoxic compounds during the thermal processing of food (for example, in meat) [30].

Parsley (*Petroselinum crispum* Hoffm) stems contain many bioactive compounds such as carotenoids, including β-carotene, neoxanthin, violaxanthin, lutein, and glycoside apiose. The leaves have a high content of vitamins (K, A, C), folate, niacin, choline, pantothenic acid and also β-carotene, lutein. Parsley is used to improve appetite and alleviate indigestion, flatulence and spasms, and may prevent stomach ulcers [31]. Stem and leaf extracts of *Petroselinum crispum* have been identified to exert antioxidant, anti-inflammatory, and antiplatelet activities, and have protective effects against hyperuricemia and hyperglycemia, brain, heart, and liver diseases [32].

It proved to inhibit the bacterial growth of *Bacillus subtilis*, involved in the pathogenesis of digestive anastomotic leak, and of the pro-inflammatory proteobacterium *Escherichia coli*, involved in Crohn's disease, whose overgrowth is facilitated by Western diets [33–35].

According to Zhang et al. [36], parsley essential oil exhibits antioxidant activity due to the compounds myristicin (phenylpropene) and apiol (phenylpropanoid). The flavonoids isorhamnetin, apigenin, quercetin, luteolin, and chrysoeriol represent the predominant compounds in cell suspension cultures of parsley [32]. Parsley apigenin (yellow color) can be used as a pigment in human and animal nutrition [1].

Wong and Kitts [37] studied the in vitro antioxidant activity by DPPH inhibition of free radicals, ion-chelating, and hydroxyl radical assays from aqueous and methanolic extracts prepared from parsley leaves and stems. The methanolic extracts from the leaves showed a significant ($p < 0.05$) radical scavenging activity, attributed to the total phenolic content, whereas the chelating activity of the ferrous ions was significantly ($p < 0.05$) higher in methanol extracts from the stem.

Phenolic compounds are the main class of plant compounds that contribute to their antioxidant capacity [38].

Due to its high nitrate content, different concentrations of parsley extract powder have been used in the formulation of sausages (V3: 1.07 g, V4: 2.14 g and V5: 4.29 g parsley extract/kg meat), in comparison, from a sensory point of view, with traditional sausages with nitrites (V1) and without nitrites (V2). The evaluators did not observe any significant differences between samples V1, V3, V4, and V5 in terms of red color intensity, although sample V2 showed intense grey color (due to lack of nitrate). Parsley extract powder did not show an influence on taste and aroma and no significant differences were found between the sausage variants. Among sausages produced with parsley extract powder, the general acceptance for sample V3 was reduced, due to the lower concentrations of parsley extract, resulting in lower levels of nitrites and lower nitrosomyochromogen contents formed. Thus, depending on the sensory parameters evaluated, samples V4 and V5, produced with 2.14 g and 4.29 g of parsley extract/kg meat, respectively, had characteristics similar to those of the traditional counterpart treated with nitrate [10].

Rosemary (*Rosmarinus officinalis* L.) leaves are rich in compounds with health-promoting properties for a variety of diseases. The main bioactive compounds are phenolic diterpenes—carnosic acid and carnosol; triterpenes—oleanolic, betulinic, and ursolic acids; flavonoids—hesperidin, homoplantaginin, cirsimaritin, genkwanin, gallocatechin, nepetrin, 6-hydroxyluteolin-7-glucoside, luteolin-3′-glucuronide, and luteolin-3′-*O*-(*O*-acetyl)-β-D-glucuronide; and typical compounds—rosmarinic acid, rosmadial, and rosmaridiphenol [39]. They confer antioxidant, anti-inflammatory, antihyperlipidemic, hepatoprotective, renoprotective, antithrombotic, antinociceptive, antidepressant, antimicrobial, and anticancer properties. In traditional medicine, *Rosmarinus officinalis* L. has been used to treat gastrointestinal, hepatic, cardiovascular, nervous, respiratory, genitourinary, and skin disorders [40,41]. In experimental models, rosemary extracts improved body weight control, total cholesterol level, and atherogenic index, cardiac remodeling after myocardial infarction, brain tolerance to artificially induced ischemia, and protection against rupture of the blood–brain barrier [42]. Anticancer properties were demonstrated in several cell-line models, for esophageal squamous cell carcinoma (KYSE30), gastric adenocarcinoma (AGS), epithelial colorectal adenocarcinoma (CaCo-2), breast adenocarcinoma (MCF-7), cervical adenocarcinoma (HeLa), lung carcinoma (A549) histiocytic lymphoma (U-937), and human melanoma (A375) [43]. Rosemary extracts also exert antifungal, antiviral, and antibacterial activities [41].

The stability, antioxidant and antimicrobial activity of rosemary essential oil introduced into meat and meat products can be improved by encapsulating it in a nanogel of chitosan and benzoic acid [44]. Krkić et al. studied the incorporation of oregano essential oil into a chitosan coating, which reduced lipid oxidation, and contributed to the formation of a smaller amount of aldehydes and superior sensory properties in dry fermented sausages [45].

Rosemary extract (0, 250, 500, and 750 mg/kg) combined with sodium nitrite (40, 80, and 120 mg/kg) was used to obtain liver pâté and to study color stability, lipid oxidation, and concentrations of ascorbic acid, α-tocopherol, and carnosic acid. Regardless of the added dose of rosemary extract, it had a significant effect in reducing lipid oxidation and maintaining high levels of antioxidants, while having no effect on color stability. The concentration of carnosic acid increases with the dose of rosemary extract. Low doses of sodium nitrite (80 mg/kg) can be used without adversely affecting color stability, forming significantly lower nitrite concentrations and slightly reduced lipid oxidation values, while the use of rosemary extract helps maintain lipid stability [46].

Rosemary and marjoram essential oils have been added in different doses to pork sausages in a study conducted by [47]. The results demonstrated the protective effects of essential oils against oxidation. In addition, the increase in TBARS values and the loss of red color was prevented and compared with samples containing synthetic antioxidants; samples with essential oil obtained similar or better results. Mohamed and Mansour evaluated the antioxidation efficiency of rosemary and marjoram essential oils (200 mg/kg) added to frozen beef patties and stored for 3 months at −18 °C. Essential oils have been shown to be effective against oxidation. The results of a sensory analysis showed that the addition of essential oils had a positive effect on the samples, being highly appreciated by evaluators [48].

According to the Food Safety and Inspection Service (FSIS) Directive 7120.1: "Safe and suitable ingredients used in the production of meat, poultry, and egg products"; rosemary extract is the most commonly used natural antioxidant in the meat industry, with it explicitly being allowed for use as a component of an antioxidant mixture [49]. The antioxidant properties of rosemary are due to the presence of phenolic diterpenes, namely carnosic acid and carnosol, which act as hydrogen donors in the reaction with free radicals [46]. In the case of sage, the antioxidant capacity is due to the presence of phenolic diterpene (epirosmanol, carnosol, and carnosic acid) [50]. Rosemary and sage extracts can provide antioxidant species in both the polar and non-polar phases of a food product. For example, carnosic acid is on the lipophilic end of the scale, and rosmarinic acid is on the hydrophilic end. Carnosic acid is a superstoichiometric antioxidant because it can act repeatedly as a reducing agent by donating hydrogen atoms through phenolic compounds [51].

Sage (*Salvia officinalis* L.) leaves are rich in terpenes, anthraquinone, and flavonoids with antioxidant, anti-inflammatory, and antimicrobial effects. In sage essential oils, the main components are camphor, 1,8-cineole, α-thujone, β-thujone, borneol, and viridiflorol [52]. In traditional medicine, *Salvia officinalis* has been used to treat mild dyspepsia, ulcers, and gout. The German Commission E has also accepted the use of *Salvia officinalis* for dyspepsia [53].

It was reported that drinking sage tea prevented the initiation phases of colon carcinogenesis in an experimental rat model [54]. Clinical trials reported memory-enhancing and antidementia benefits in healthy adults or patients with Alzheimer's disease [55], and hypoglycemic and hypolipemic effects of *Salvia officinalis* leaves in patients with diabetes and hyperlipidemia and in healthy volunteers [53].

Sage may produces several types of phenolic species as opposed to rosemary, especially in the production of flavonoids and other phenolic derivatives, although rosemary produces higher amounts of carnosic acid and other diterpenoids related to it [51]. In a study by Kontogianni et al., they showed that rosemary extract is twice as rich in diterpornoid and phenolic compounds and contains about 2.7 times more carnosic acid than sage extract [56].

Dill (*Anethum graveolens* L.) seeds used in meat products and derivatives are rich in essential oil, of which the major compounds are carvone, limonene, and camphor, characterized by important antioxidant activity [57]. Antioxidant activity also characterizes the flavonoids quercetin and isoharmentin isolated from *Anethum graveolens* L. seeds, which can help to prevent peptic ulcers. This effect has been verified in experimental models in which aqueous and ethanolic *Anethum graveolens* L. extracts had mucosal protective and antisecretory effects, similarly to high-dose sucralfate [58].

The essential oils found in seeds are carminative, improve appetite, aid digestion, and relieve intestinal spasms. D-limonene is a monoterpene that dissolves cholesterol-containing gallstones. It is chemopreventive and has chemotherapeutic activities [59]. In experimental models, dill seed extracts suppressed hyperlipidemia induced by a high-fat diet [60] and had inhibitory effects on hepatic carcinoma cells [61].

Extracts from dill (*Anethum graveolens*) obtained in organic and conventional agriculture were prepared in n-hexane, dichloromethane (CH_2Cl_2), ethyl acetate (EtOAc) and ethanol (EtOH), and the radical scavenging activity (RSA) at 2000 μg mL^{-1} has been studied by DPPH (2,2-Diphenyl-1-picrylhydrazyl), DMPD+ (N,Ndimethyl-p-phenylendiamine), and NO (nitric oxide) methods. Ethanol extracts (both conventional and organic agriculture) had better inhibitory effects, and the NO radical scavenging activity was particularly noted (78.49 ± 1.86% for conventional agriculture and 71.86 ± 5.41% for organic agriculture). No significant differences were observed between organic and conventional agriculture extracts in the RSA tests. Ferric-ion-chelating capacity and phosphomolybdenum-reducing antioxidant power (PRAP) assays were also studied; dichloromethane (CH_2Cl_2) extracts had the greatest ferric ion chelation effect (74.34 ± 1.40%), and PRAP values from both extracts had better values than the rest of the samples [62].

Oregano (*Origanum vulgare* L.) leaves are rich in essential oils. The volatile oil contains phenolic compounds, monoterpenes and sesquiterpenes: thymol, carvacrol, *p*-cymene, γ-terpinene, and linalool [63], with antioxidant, anti-inflammatory, and antimicrobial activity. In traditional medicine, oregano has been used for gastrointestinal disorders—indigestions, stomachache, and diarrhea; respiratory diseases—asthma and bronchitis; menstrual disorders; and diabetes, due to its anti-bacterial and anti-inflammatory activity [64–67].

A remarkable property of oregano essential oils is their antiproliferative activity on adenocarcinoma gastric cell line [68]. In case–control studies, gastric cancer was associated with red and processed meat consumption [69]. *Origanum vulgare* essential oils demonstrated inhibitory effects on the growth of carbapenem-resistant Gram-negative bacteria [70], and antibacterial activity and synergistic effect with polymyxin B against multidrug-resistant *Acinetobacter baumannii* [71], whose development is favored by extensive antibiotic utilization, including animal treatments [72,73].

Oregano has a high content of antioxidants, according to a study conducted by Zheng and Wang, which makes it suitable for use as a natural antioxidant. Antioxidant compounds are phenolic acids and flavonoids, such as caffeic acid, rosmarinic acid, hispidulin, and apigenin, as well as carvacrol and thymol, components of the essential oil [74].

Fasseas et al. evaluated the antioxidant activity of meat treated with oregano and sage essential oils extracted by hydrodistillation. In this regard, the minced pork and beef were formulated in three samples as follows: homogenization with 3% (*w*/*w*) of either oregano essential oil or sage essential oil, and a control sample which did not contain essential oils. The samples thus obtained were stored at 4 °C in raw and cooked states (85 °C, 30 min), and the antioxidant activity was evaluated at 1, 4, 8, and 12 days of storage. Essential oils have led to a decrease in lipid oxidation, the role of antioxidants being affected by meat proteins and was significantly more important in cooked meat [75].

Plant extracts and essential oils including thyme, oregano (rich in thymol and carvacrol), rosemary, and sage are used to prevent the oxidation of meat products, in encapsulated form or in edible films, due to their high solubility, and for flavoring, due to their organoleptic properties [76].

The effect of oregano essential oil used in the formulation of an active coating used on fresh pork meat was studied. The essential oil was used as free oil, nanoemulsified or microencapsulated. All formulated samples showed a delay in the oxidation of lipids and oxymyoglobin, and the sensory profile was more appreciated as opposed to the control sample which did not contain oregano essential oil [44].

Basil (*Ocimum basilicum* L.) leaves contain many antioxidant and anti-inflammatory flavonoids such as quercetin, quercetin-3-*O*-diglycoside, querce-tin-3-*O*-β-D-galactoside, quercetin-3-*O*-β-D-glucoside, quercetin-3-*O*-β-D-glucoside-2″-gallate, quercetin-3-*O*-(2″-*O*-galloyl)-rutinoside, querce-tin-3-*O*-α-L-rhamnoside, isoquercetrin, kaempferol; carotenoids—β-carotene, β-cryptoxanthin, and lutein–zeaxanthin; polyphenols—rosmarinic acid, and chicoric acid (dicaffeoyltartaric acid); coumarin, aesculetin, and *p*-coumaric acid [77]. In traditional medicine, *Ocimum basilicum* has been used for the treatment of digestive disorders and demonstrates carminative, stimulant, antispasmodic, antidiarrheal, antibacterial, and anthelmintic effects [31]. It has also been used in treating vomiting, flatulence, dyspepsia, and gastritis [78]. *Ocimum basilicum* leaves contain caffeic acid, which demonstrates antioxidative and cancer chemopreventive properties [79]. *Ocimum basilicum* essential oils exhibited cytotoxic activity against human liver hepatocellular carcinoma cell lines (HEpG2) and nasopharyngeal cancer cell line (KB) [78]. In experimental models, *Ocimum basilicum* demonstrated anti-hyperglycemic potential, and antioxidant and nephroprotective effects in diabetic disease [80–82].

The antioxidant compounds in basil extracts with a role in antiradical activity are chlorogenic, *p*-hydroxybenzoic, caffeic, vanillic, and rosmarinic acids, as well as apigenin, quercetin, and rutin. Teofilović et al. performed various extractions with mixtures of ethanol–water (30%, 40%, 50%, 60%, 96% v/v), concentrated methanol (95% v/v), water (in presence and absence of light), dichloromethane, chloroform, and hexane over different periods of time (10 and 30 min), which exhibited antioxidant activity by the DPPH method with IC_{50} values between 0.22 ± 0.01 and 12.99 ± 0.87 g/mL for polar solvents and from 12.12 ± 0.54 to 20.49 ± 1.54 g/mL for non-polar solvents. Increasing the extraction time and polarity of the solvent improve the quality of the extracts in terms of phenolic compounds and antioxidant capacity [83].

Basil essential oil was added in various concentrations (0.062%, 0.125%, and 0.25%) to a beef burger to evaluate its natural antioxidant effectiveness. The results showed that the essential oil decreased the rate of lipid oxidation, and the effectiveness did not depend on the concentration [13].

Marjoram (*Origanum majorana* L.) leaf essential oils have flavonoids and terpenoids as the main active compounds [42,84]. These are excellently summarized in a review published by Bina [85]. Monoterpene hydrocarbons are represented by α- and β-pinene, α- and β-phellandrene, camphene, sabinene, limonene, ρ-cymene, β-ocimene, γ-terpinene, α-terpinene, terpinolene, carvone, and citronellol. Thymol, carvacrol, and linalool are other monoterpene compounds. Sweet marjoram essential oil contains phenolic compounds such as rosmarinic acid, sinapic acid, vanillic acid, ferulic acid, caffeic acid, and coumarinic acid, and phenolic glycosides such as arbutin, methyl arbutin, vitexin, and orientinthymonin [85]. Flavonoids such as hesperetin, kaempferol, and luteolin were also found in marjoram extracts and have vasoprotective effects, together with carvacrol and thymol [86]. Marjoram is traditionally used to treat respiratory and gastrointestinal diseases, high blood pressure, dysrhythmia, pains, and fatigue. Marjoram extracts demonstrated antioxidant, anti-inflammatory, anti-hyperglycemic, hypouricemic, anticancer, gastro-, nephro-, and hepatoprotective activity [85]. Acetylcholinesterase and tyrosinase inhibitory activities were also demonstrated, and support the antineurodegenerative effects of marjoram [87]. Essential oils and extracts from *O. majorana* exhibit antiparasitic, antifungal and antimicrobial activity against the Gram-positive species *Staphylococcus aureus*, *Enterococcus faecalis*, and *Streptococcus dysgalactiae*, and the Gram-negative species *Klebsiella pneumoniae* and *Pseudomonas aeruginosa* [88–90].

Marjoram essential oil has antioxidant properties through its ability to inhibit hydroxyl radicals. The major compounds of the oil are terpinen-4-ol (21.3%), *trans*-sabinene hydrate (15.5%), γ-terpinene (14.0%), and α-terpinene (8.9%). At a concentration of 0.05%, marjoram essential oil inhibited the formation of conjugated dienes by 50% and the generation of oxidized by-products of linoleic acid by 79.85% through its addition to an emulsion system with linoleic acid [91].

Mint (*Mentha piperita* L.) leaf essential oils mainly contain oxygenated monoterpenes, monoterpene hydrocarbons, sesquiterpene hydrocarbons, and oxygenated sesquiterpenes [92]. Quantitatively, the most abundant constituents are menthol and menthone. Menthofuran, menthyl acetate, iso-menthone 1,8-cineole, and the toxicity of pulegone should be remarked [93]. The traditional use of mint for treating fevers, colds, digestive diseases, infections, and throat inflammation has been supported by experimental studies. Pharmacological activities of mint demonstrated to date include antioxidant, anti-inflammatory, anticancer, antidiabetic, hepatoprotective, neuroprotective, and radioprotective activity [94]. In healthy adults, essential oil rich in menthol/menthone attenuated mental fatigue associated with extended cognitive task performance [95]. Antimicrobial, antiviral, antifungal, biopesticidal, and larvicidal activity has also been reported.

Mint extracts contain a significant number of phenolic compounds, and thus exert an important antioxidant activity. Their antioxidant activity can be compared with that of synthetic antioxidants [96].

In their study, Kanatt et al. treated lamb pulp with mint extract before it was irradiated (2.5 kGy). An amount of 0.05 g/100 g mint extract inhibited oxidation to some extent (0.6 mg MDA/kg as opposed to using a 0.1 g/100 g extract, which led to inhibition of 50% (0.4 mg MDA/kg) [97]. In another study, Biswas et al. treated ground pork meat with mint extract, and it obtained a good color stability compared with samples obtained with sodium nitrite [98].

Tarragon (*Artemisia dracunculus* L.) leaf essential oils are rich in phytochemicals including coumarins, isocoumarins, monoterpenoids, sesquiterpenoids, flavonoids, polyacetylenes, and alkaloids. Tarragon increases bile and gastric acid production, stimulates digestion, and has beneficial effects on gastritis [99]. Essential oil concentrated in coumarin derivatives showed remarkable anticoagulant activity, inducing a therapeutic value (2.34) for the international normalized ratio (INR) in vitro [100]. In muscle cell cultures derived from lean, overweight, and diabetic–obese subjects, bioactive compounds of *Artemisia dracunculus* L. improved insulin sensitivity [101]. Essential oils also exhibited strong antifungal activity [102].

The antioxidant capacity of tarragon (*Artemisia dracunculus*) essential oil (0.01–0.9%) was studied by Behbahani et al. and the results showed that a concentration of 0.9% has an antioxidant activity of 78.87%, similar to that of the synthetic antioxidant BHT (butylated hydroxytoluene) [102].

Nimse and Pal present carotenoids, antioxidant vitamins, hydroxycinnamic acids, flavonoids and terpenes as antioxidant compounds that can help prevent oxidation in meat and meat products [103]. The phenolic content of plant-derived materials has the most significant potential in terms of the antioxidant and antimicrobial activity [76].

Coriander (*Coriandrum sativum* L.) leaves have high contents of vitamin C, vitamin A, vitamin K, iron, manganese, thiamine, zinc, β-carotene, and anthocyanins, and are used for the treatment of iron and vitamin deficiencies or as potent antioxidants. Fruits and leaves of *Coriandrum sativum* are traditionally used for digestive diseases, nausea, vomiting, indigestion, and against worms [104]. In experimental models, coriander seeds prevented gastric mucosal lesions induced by ethanol due to the protective layer formed by its hydrophobic compounds and free radical scavenging activity of its antioxidant constituents such as flavonoids, coumarins, catechins, and terpenes [105]. *Coriandrum sativum* L. leaf essential oils contain natural antimicrobial compounds that can act against *Candida* spp. [106] and *Campylobacter jejuni* found in beef and chicken meat [21]. Previous studies have demonstrated the antioxidant and neuroprotective effect of *Coriandrum sativum* L. extracts on brain [107], the decrease in brain cholinesterase activity and serum total cholesterol levels, and memory improvement [108]. Extracts from the leaves and stems of *Coriandrum sativum* L. exhibited significant antihyperglycemic activity, and seed extracts normalized glycemia and decreased the elevated levels of insulin [109].

Šojić et al. [110] effectively investigated the addition of coriander essential oil (0.075–0.150 µL/g) to pork sausages which also contained different levels of sodium nitrite (0, 50 and 100 mg/kg). In addition to nitrite, coriander essential oil contributes to lower lipid oxidation due to its antioxidant potential due to terpenoid compounds. The essential oil contains linalool (835.2 mg/g), camphor (32.9 mg/g), γ-terpinene (32.8 mg/g), geraniol (16 mg/g), and (+)-limonene (6.2 mg/g) [111].

Bay (*Laurus nobilis* L.) leaf essential oils contain monoterpenes and monoterpenoids, sesquiterpenes and sesquiterpenoids, diterpenoids, phenyl propene derivatives, alcohols, carbonyls, and esters [112]. *Laurus nobilis* L. is used in the treatment of cancer, gastrointestinal disorders, epilepsy, rheumatic conditions, and several infectious diseases [113]. In an experimental model of ulcers induced by ethanol, bay leaf extracts demonstrated gastric mucosal protection correlated with antioxidant activity [114].

The antimicrobial activity of essential oils tested against *Escherichia coli*, *Staphylococcus aureus*, *Enterococcus faecalis*, *Pseudomonas aeruginosa*, and *Candida albicans* exhibited inhibitory effects and antioxidant activity [112].

Bay leaf essential oil has antioxidant and antibacterial properties in meat due to oxygenated monoterpenes and phenolic compounds, according to the study conducted by Ramos et al. [115]. Bay leaf essential oil has been used to treat packaged chicken in a microaerobic atmosphere where it has reduced oxidation and prolongs the smell of fresh meat, making it suitable for use as a natural preservative. Bornyl acetate, 1,8-cineole, β-myrcene, and carvacrol represents the principal components from bay leaf essential oil [116].

Thyme (*Thymbra capitata* (L.) Cav) leaf essential oils contain mainly monoterpenes, generally 10% carvacrol and 50% thymol, but also linalool, α-terpineol, camphor, caryophyllene, and γ-terpinene [117]. Flavonols (quercetin-7-*O*-glucoside), flavanones (naringenin) and flavones (apigenin), phenolic acids (*p*-coumaric, caffeic, rosmarinic, cinnamic, carnosic, ferulic, quinic, and caffeoylquinic acids), saponins, steroids, alkaloids, and tannins have also been described in thyme extracts [117–119]. In traditional medicine, thyme has been used as a sedative, a carminative, an additive for baths, or an infusion for the treatment of skin diseases [117,118]. Thyme extract might be an effective treatment of chronic respiratory diseases accompanied by the inflammation and hypersecretion of mucus [120]. Previous studies have reported that thymol demonstrates anti-inflammatory, anti-carcinogenic, and immunomodulatory properties, decreased serum lipids, visceral fat accumulation, and reduced blood pressure in experimental models [121,122]. Thyme essential oils have exhibited antifungal and antibacterial bioactivity against both food spoilage microflora and pathogenic microflora in vitro, including *B. subtilis*, *S. aureus*, *S. enteritidis*, and *P. aeruginosa* [123]. Thyme essential oils demonstrated antimicrobial activity against *Listeria monocytogenes* in vivo [124].

Due to the carvacrol present in the composition of thyme essential oil, it is distinguished by its antioxidant capacity. Thus, through their study on chicken breast in which thyme essential oil was added in a proportion of 0.5%, Fratianni et al. demonstrated that it reduced radical formation and lipid peroxidation and prolonged the shelf life of products [125]. In another study, Zengin and Baysal added thyme essential oil to minced beef that was stored for 9 days at 4 °C. Thyme essential oil showed a delaying effect in the oxidation process of lipids and color, and the sensory quality of the product was not affected by this addition [126].

4. Essential Oils and Aroma Compounds of Spicy and Aromatic Plants

Chemically, essential oils are a complex mixture of numerous bioactive chemical components, such as terpenes, terpenoids, and phenolics. Essential oils are synthesized by almost all plant organs, particularly the flowers, buds, leaves, seeds, stems, roots, and fruits [127]. Moreover, essential oils exhibit a very characteristic odor, and are therefore responsible for the specific scents that aromatic plants emit. These essential oils can be stored in epidermal cells, cavities, secretory cells of glandular trichomes. Numerous essential oils have the potential to be used as a food preservative for meat and meat products [128–133]. It should be highlighted that essential oils are generally accepted by consumers due to their high volatility and biodegradable nature.

Rosemary. Oxygenated monoterpenes (74.1%), represented mainly by 1,8-cineole (33.1%), camphor (18%), and borneol (7.95%), are the major terpenes of the Tunisian rosemary essential oil (Table 2). Monoterpene hydrocarbons constitute 21.6% of the oil, and α-pinene (10.16%) is the major compound of this class. The amount of oxygenated sesquiterpenes, represented only by caryophyllene oxide, was low (0.62%) [134].

According to the literature [44,135–138], the main primary components of the oil are: 1,8-cineole, α-pinene, camphor, verbenone, and borneol, whereas the secondary components are terpinen-4-ol, α-terpineol, β-caryophyllene, 3-octanol, geranyl acetate, and linalyl acetate. *Rosmarinus officinalis* L. volatile oil can be distinguished according to its 1,8-cineole, α-pinene, and camphor content. Additionally, certain essential oil compositions are dominated by myrcene [139].

The volatile fraction of *R. officinalis* differs for many of the chemotypes growing in the countries listed in Table 2, with regard to compounds, in the genus of components and their relative quantity. The observed differences may probably be due to different environmental and different chemotypes and the nutritional status of the plants, as well as other factors that can influence the oil composition. It is of interest to note the presence of *p*-cymene (44.42%) in very high percentages, which was distinctive of *R. officinalis* [140]. Monoterpenes constituted the major compounds of the oil (86.3%), whereas sesquiterpenes amounted to 11% [141].

Mint. The essential oil of *M. pullegium* is rich in pulegone (35.1%) and piperitenone (27.4%) (Table 2). However, none of these compounds were present in the essential oil of the other Mentha species, i.e., mint, whose main component is carvone (75.9%). Pulegone (48%) and menthone (41%) are described in the essential oil of the *M. pullegium* species [135]. According to Erich et al., the chemical composition of peppermint oil, as detailed in Table 2, is dominated by menthol (40.7%) and menthone (23.4%) [142]. The odor perceptions of peppermint essential oil include green, herbaceous, bitter, mint, or fresh. The amount of peppermint essential oil is influenced by geographical area, ripening time, or soil type [143].

Thyme. The major component identified from wild thyme was carvacrol (56.0%) [135]. Carvacrol (55.1%) and geraniol (43%) were the main components of the oils from *Thymus x citriodorus* "Archer's Gold" and Thymus x citriodorus, respectively (Table 2). Among the other constituents, *p*-cymol, β-caryophyllene, geranial, limonene, and γ-terpinene were also characteristic of the oils, but smaller amounts were present. Thymol is the main component of oils from *Thymus vulgaris* and *Thymus serpyllum* (37.1% and 17%, respectively) [144].

Coriandrum. Essential oils from the fresh herb *Coriandrum sativum* L. accumulate during the growth of the plant. According to Wei et al., it is recommended to harvest the vegetative part prior to flowering because (*E*)-2-decenal, a potential irritant, is present at higher percentages in the preflowering and full flowering stages [145].

There are significant differences in the essential oil types from different parts of coriander. (*E*)-2-decenal is the dominant constituent in essential oils from coriander leaves: 31.28% and 61.86% of the essential oils from coriander stems [145]. Linalool was found to be the main constituent of essential oils from fully mature coriander fruits, seeds, and pericarps. Linalool and citronellol are the main components of coriander inflorescence essential oils from lower latitudes. The young vegetative organs in the seedling stages not only have a grassy odor, peculiar to coriander, but also a fresh green odor, highly suitable for seasoning [146].

Sage it is one of the most appreciated herbs due to its rich essential oils, which have antimicrobial, antifungal, and antimutagenic properties [128]. More than 120 components have been characterized in the essential oil prepared from aerial parts of *S. officinalis*. The main components of the oil include borneol, camphor, caryophyllene, cineole, elemene, humulene, ledene, pinene, and thujone [53].

Basil. Oxygenated monoterpenes and phenylpropanoids are the main compounds of the *Ocimum* genus. In different *O. basilicum* cultivars and chemotypes, linalool, eugenol, methyl chavicol, methyl cinnamate, methyl eugenol, and geraniol have been reported as the major components [147].

Parsley. In terms of essential oils, the major compound in parsley's essential oils is myristicin, which has spicy, warm, and balsamic odors. The combined cleaning process with drying at different temperatures provided a greater reduction in the microorganisms in relation to separate processes. Drying did not change the oil yield in relation to the fresh plant [148].

Tarragon. According to Sobieszczanska et al., tarragon essential oil and its major compounds act against food-associated *Pseudomonas* spp. Tarragon essential oils are rich in other compounds, particularly methyl chavicol (tarragon and basil essential oils [149].

Origanum essential oils, together with sage, rosemary, marjoram, and thyme essential oils, are mainly composed of oxygenated monoterpenes (>40%) [150]. Generally, the *Origanum* species is characterized by the presence of two major biochemically related groups of compounds (aromatic monoterpenes such as *p*-cymene, thymol, carvacrol, their precursor γ-terpinene, and their derivatives; thujanes, such as sabinene, sabinene hydrate, and their derivatives) [151]. Thymol and carvacrol, which are present in high amounts in its essential oils (78–82%), are generally responsible for its antioxidant properties [128].

Marjoram. The results of marjoram essential oil analysis by Dimitra et al. gave a large number of constituents. Among them, 3-thujene (2.8%), γ-myrcene (3.8%), 2-carene (7.8%), 2-ethyl-mxylene (5.2%), 3-carene (10.4%), terpinen-4-ol (7.8%), sabinene hydrate (6.0%), R-terpineol (4.2%), and thymol (14%) were detected. Two chemotypes of *O. majorana* are reported in the literature: the cis-sabinene hydrate/tepinen-4-ol chemotype and the carvacrol/thymol chemotype [151].

Bay. The predominant flavor compound is 46% eucalyptol; essential oils of bay leaf induced human aryl hydrocarbon receptor activity by threefold, and its major constituent eucalyptol (46%) was inactive [152].

Dill. Terpenes were the most abundant volatiles detected in dill essential oils. Dill oil had a relatively limitary chemical profile, approximately equal volumes of carvone and D-limonene. Together, these accounted for 97.5% of the compounds identified by gas chromatography/mass spectroscopy [153].

Table 2. Variability in the chemical composition of EOs and aroma compounds of spices and aromatic plants used to prepare meat and meat alternatives.

	Origin	Chemotype	Odor Perception [a]	References
Rosemary				
Rosmarinus officinalis L.	Mexico	14.1% α-Pinene, 11.5% camphene, 12.0% β-pinene, 7.9% α-phellandrene, 8.6% 1,8-cineole, 3.4% 2-bornanone, 8.7% camphor	Pine, camphor, turpentine, resin, turpentine, turpentine, mint, spice, sweet	[154]
	Australia, USA, South Africa, Kenya, Nepal, and Yemen	13.5%–37.7% α-pinene, 16.1%–29.3% 1,8-cineole, 0.8%–16.9% verbenone, 2.1%–6.9% (−)-borneol, 0.7%–7.0% camphor, 1.6%–4.4% racemic limonene.	Pine, mint, sweet, camphor	[137]
	Brazil	26.0% Camphor, 22.1% 1,8-cineol, 12.4% myrcene, 11.5% α-pinene	Camphor, pine, turpentine, mint, sweet, balsamic, must, spice	[136]
	Portugal	1.2% α-humulene, 7.2% α-terpineol, 35.4% verbenone	Wood, oil, anise, mint	[135]
	Portugal	16.6–29.5% Myrcene, 8.3–14.5% 1,8-cineol, 14.3–23.1% camphor	Mint, sweet, camphor, balsamic, must, spice	[138]
	Morocco	37.4% α-pinene, 41–53% camphor, and 58–63% 1,8-cineol	Pine, turpentine, mint, sweet, camphor	[155]
	Tunisia	20–46% 1,8-cineol, 8.5–30.2% camphor, 6.5–13% α-pinene, 4–25% borneol	Pine, turpentine, mint, sweet, camphor	[134]
	Algeria	48.9% Camphor		[139]
	Egypt	52.8% 1,8-cineol, 11.9% camphor, 10.2% α-pinene, 7.5% borneol	Mint, sweet, camphor	[156]
	Lebanon	19.1–25.1% 1,8-cineol, 18.8–38.5% α-pinene	Pine, turpentine, mint, sweet	[157]

Table 2. *Cont.*

	Origin	Chemotype	Odor Perception [a]	References
	Turkey	44.02% p-Cymene, 20.5% linalool, 16.62% γ-terpinene, 2.64% 1,8-cineol	Solvent, gasoline, citrus, turpentine, mint, sweet, flower, lavender	[140]
	Greece	24.1% α-Pinene, 14.9% camphor, 9.3% 1,8-cineol, 8.9% camphene	Pine, turpentine, camphor, mint, sweet, camphor	[141]
	Sardinia (Italy)	23% α-Pinene, 16% borneol, 9.4% verbenone, 10.4% bornyl acetate	Pine, turpentine	[158]
	Spain	18.2% Eucalyptol, 35.5% (−)-Camphor, 13.4% (−)-Bornylacetate	Camphor	[150]
	Spain, Morocco, and Tunisia	5–21% camphor, 15-55% 1,8-cineole, 9–26% pinene, 1.5–5.0% borneol, 2.5–12.0% camphene, 1.5–5.0% limonene.	Pine, turpentine, camphor, mint, sweet	[40]
Mint				
Mentha spicata	Portugal	75.9% Carvone	Mint	[135]
Mentha pullegium	Portugal	35.1% Pulegone, 27.4% Piperitone	Aromatic, minty, green, herbaceous, bitter, mint, fresh	
Mentha viridis	Morocco	37.26% carvone, 11.82% 1,8-cineole, 08.72% Terpinen-4-ol(leaves)	Citrus; herbaceous; fruity; sweet; vanilla; minty; pepper; spicy; woody, turpentine, nutmeg, must	[159]
Mentha x piperita L.	USA	40.7% menthol, 23.4% menthone	Cool-minty, minty	[142]
Thyme				
Thymus serpyllum	Portugal	56.0% carvacrol, 4.9% α-terpineol, 2.7% veridiflorol	Oil, anise, mint, sweet, green, herbal, fruity, tropical, minty	[135]
Thymus vulgaris	Hungary	37.1% thymol, 3.1% carvacrol	Herb, spicy	[144]
Thymus serpyllum		17% thymol, 2.3% carvacrol	Herb, spicy	
Thymus x citriodorus		0.8% thymol, 6.1% carvacrol, 43% geraniol	Herb, spicy, rose, geranium	
Thymus x citriodorus "Archer's Gold"		0.7% thymol, 55.1% carvacrol	Herb, spicy	
Thymus mastichina	Spain	75.4% m-thymol, 5.4% carvacrol	Herb, spicy	[150]
Thymus vulgaris	Greece	4.3% γ-terpinene, 23.5% p-cymene, 2.2% carvacrol 63.6% thymol	Gasoline, turpentine, solvent, citrus	[151]
Coriander				
Coriandrum sativum L.	North India (Haldwani)	62.1% linalool, 7.3% (2e)-dodecenal, 4.1% n-dodecanal, 4.1% α-pinene (inflorescence eo)	Flower, lavender, green, fat, sweet, pine, turpentine	[160]
	Tunisia (Korba)	86.1%, 91.1% and 24.6% linalool (fruit, seed, and pericarp)	Flower, lavender	[161]
	Seoul, Korea	23.11% cyclododecanol, 17.86% tetradecanal, 9.93% 2-dodecenal, 7.24% 1-decanol, 6.85% 13-tetradecenal, 6.54% 1-dodecanol, 5.16% dodecanal, 2.28% 1-undecanol, and 2.33% decanal (leaves)	Mandarin, fat, soap, orange peel, tallow	[162]
	Austria	60.5 % δ^3-carene, 18.2% γ-terpinene,	Orange peel, gasoline, turpentine, lemon, resin	[150]
Salvia				
Salvia officinalis	Mexico	12.2% eucalyptol, 28.7% (−)-camphor, 12.6% (−)-bornylacetate	Citrus, herbaceous; fruity, sweet, vanilla, minty, pepper, spicy, woody, camphor	[150]
	Italy	20.16% α-thujone, 14.04% 1,8-cineole, 10.09% β-pinene (flower)	Pine, resin, turpentine, mint, sweet	[163]

Table 2. *Cont.*

	Origin	Chemotype	Odor Perception [a]	References
Basil				
Ocimum basilicum	The Island of Comoro	77.9% methylchavicol	Licorice, anise	[150]
Ocimum basilicum cv. Keshkeni luvelu	Iran	linalool, 1,8-cineole, tau-muurolol, and α-cadinol (major compounds)	Flower, lavender, mint, sweet	[147]
Parsley				
Petroselinum sativum	USA	45.1% 4-methoxy-6-(2-propenyl)-1.3-benzodioxole	Spice, warm, balsamic	[150]
Petroselinum crispum Mill.	Brazil	apiole, myristicin (major compounds)	Spice, warm, balsamic	[148]
Tarragon				
Artemisia dracunculus	Spain	92.4% methylchavicol	Licorice, anise, clove, spice, mint, turpentine	[150]
	Spain	24.5% methyl eugenol, 19.3% β-phellandrene	Clove, spice, mint, turpentine	
Origanum				
Thymbra capitata (L.) Cav.		78.4% carvacrol, 10.9% m-thymol	Herb, spicy	[150]
Oroganum dictamus	Greece	12.7% γ-terpinene, 9.9% p-cymene, 7.8% carvacrol, 63.3% thymol	Gasoline, turpentine, solvent, citrus	[151]
Origanum vulgare L.		46.84% thymol, 12.88% γ-terpinene	Gasoline, turpentine	[163]
Marjoram				
Organum majorana	Greece	2.8% 3-thujene, 3.8% β-myrcene, 7.8% 2-carene, 5.2% 2-ethyl-mxylene, 10.4% 3-carene, 7.8% terpinen-4-ol, 6.0% sabinene hydrate, 4.2% β-terpineol, 14% thymol.	Balsamic, must, spice, lemon, resin, wood, green, herb	[151]
Bay				
Laurus nobilis	USA	46% eucalyptol	Citrus, herbaceous, fruity, sweet, vanilla, minty, pepper, spicy, woody	[152]
Dill				
Anethum graveolens		40% limonene, 44% caraway, 25% spearmint, 9% star anise.	Lemon, orange	[152]

[a] Odor descriptions according to the Flavornet (www.flavornet.org(accessed on 3 January 2022) and Pherobase databases (www.pherobase.com(accessed on 3 January 2022)).

5. Conclusions

The increasing interest of consumers for most natural foods, natural flavors, natural antioxidants, natural antimicrobial substances is often linked to their bioactive compounds; spices and aromatic plants are some of the richest sources of phytochemicals with demonstrated biological activities. The most commonly used spice and aromatic plants in the meat and meat analogues industry are parsley, dill, basil, oregano, sage, coriander, rosemary, marjoram, tarragon, bay, thyme, and mint. As shown in this paper, they improve meat preparation in terms of flavor, as well as antimicrobial and antioxidant activity. Due to their bioactive compounds, they also have beneficial effects on the consumer health. Although studies of the effects of these herbs and spices on meat analogues are limited, their use is highly recommended due to the flavor improvements, as well as consumers health benefits.

Author Contributions: Conceptualization, R.A.M., V.M. and A.E.M.; writing—original draft preparation, R.A.M., C.C.M., A.E.T., A.P., G.S.M. and R.A.U.; writing—review and editing, R.A.M., V.M. and A.E.M.; visualization, R.A.M., V.M. and A.E.M.; supervision, R.A.M., V.M. and A.E.M.; funding acquisition, V.M. All authors have read and agreed to the published version of the manuscript.

Funding: This study was supported by a grant of the Ministry of Research, Innovation and Digitization, CNCS/CCCDI—UEFISCDI, project number PN-III-P2-2.1-PED-2019-5346, within PNCDI III, and the publication was supported by funds from the National Research Development Projects to finance excellence (PFE)-14/2022-2024 granted by the Romanian Ministry of Research and Innovation.

Institutional Review Board Statement: Not applicable.

Informed Consent Statement: Not applicable.

Data Availability Statement: Not applicable.

Conflicts of Interest: The authors declare no conflict of interest.

References

1. Christaki, E.; Giannenas, I.; Bonos, E.; Florou-Paneri, P. *Innovative Uses of Aromatic Plants as Natural Supplements in Nutrition*; Academic Press: New York, NY, USA, 2020; pp. 19–34. [CrossRef]
2. Zeng, Z.; Zhang, S.; Wang, H.; Piao, X. Essential oil and aromatic plants as feed additives in non-ruminant nutrition: A review. *J. Anim. Sci. Biotechnol.* **2015**, *6*, 7. [CrossRef] [PubMed]
3. Gadde, U.; Kim, W.H.; Oh, S.T.; Lillehoj, H.S. Alternatives to antibiotics for maximizing growth performance and feed efficiency in poultry: A review. *Anim. Health Res. Rev.* **2017**, *18*, 26–45. [CrossRef] [PubMed]
4. Christaki, E.; Bonos, E.; Giannenas, I.; Florou-Paneri, P. Aromatic Plants as a Source of Bioactive Compounds. *Agriculture* **2012**, *2*, 228–243. [CrossRef]
5. Lorenzo, J.M.; Domínguez, R.; Agregán, R.; Gonçalves, A. Effect of fat replacement by olive oil on the physico-chemical properties, fatty acids, cholesterol and tocopherol content of pâté. *Grasas Aceites* **2016**, *67*, 133. [CrossRef]
6. Fernandes, R.P.P.; Trindade, M.A.; Tonin, F.G.; Pugine, S.M.P.; Lima, C.G.; Lorenzo, J.M.; de Melo, M.P. Evaluation of oxidative stability of lamb burger with Origanum vulgare extract. *Food Chem.* **2017**, *233*, 101–109. [CrossRef]
7. Hashemi, S.M.B.; Nikmaram, N.; Esteghlal, S.; Khaneghah, A.M.; Niakousari, M.; Barba, F.J.; Roohinejad, S.; Koubaa, M. Efficiency of Ohmic assisted hydrodistillation for the extraction of essential oil from oregano (Origanum vulgare subsp. viride) spices. *Innov. Food Sci. Emerg. Technol.* **2017**, *41*, 172–178. [CrossRef]
8. Sun, C.; Ge, J.; He, J.; Gan, R.; Fang, Y. Processing, Quality, Safety, and Acceptance of Meat Analogue Products. *Engineering* **2020**, *7*, 674–678. [CrossRef]
9. Boukid, F. Plant-based meat analogues: From niche to mainstream. *Eur. Food Res. Technol.* **2021**, *247*, 297–308. [CrossRef]
10. Riel, G.; Boulaaba, A.; Popp, J.; Klein, G. Effects of parsley extract powder as an alternative for the direct addition of sodium nitrite in the production of mortadella-type sausages—Impact on microbiological, physicochemical and sensory aspects. *Meat Sci.* **2017**, *131*, 166–175. [CrossRef]
11. Hajji, M.; Falcimaigne-Gordin, A.; Ksouda, G.; Merlier, F.; Thomasset, B.; Nasri, M. A water-soluble polysaccharide from Anethum graveolens seeds: Structural characterization, antioxidant activity and potential use as meat preservative. *Int. J. Biol. Macromol.* **2021**, *167*, 516–527. [CrossRef]
12. Kocić-Tanackov, S.; Dimić, G.; Ðerić, N.; Mojović, L.; Tomović, V.; Šojić, B.; Ðukić-Vuković, A.; Pejin, J. Growth control of molds isolated from smoked fermented sausages using basil and caraway essential oils, in vitro and in vivo. *LWT* **2020**, *123*, 109095. [CrossRef]
13. Chaleshtori, R.; Rokni, N.; Rafieian-kopaei, M.; Drees, F.; Salehi, E. Antioxidant and Antibacterial Activity of Basil (*Ocimum basilicum* L.) Essential Oil in Beef Burger. *J. Agric. Sci. Technol.* **2015**, *17*, 817–826.
14. Busatta, C.; Mossi, A.; Rodrigues, M.R.; Cansian, R.; Oliveira, J. Evaluation of Origanum vulgare essential oil as antimicrobial agent in sausage. *Braz. J. Microbiol.* **2007**, *38*, 610–616. [CrossRef]
15. Marques, J.; Funck, G.; Volcao, L.; Schneid Kroning, I.; Silva, W.; Fiorentini, A.; Ribeiro, G. Antimicrobial activity of essential oils of *Origanum vulgare* L. and *Origanum majorana* L. against Staphylococcus aureus isolated from poultry meat. *Ind. Crop. Prod.* **2015**, *77*, 444–450. [CrossRef]
16. Zhang, L.; Lin, Y.H.; Leng, X.J.; Huang, M.; Zhou, G.H. Effect of sage (*Salvia officinalis*) on the oxidative stability of Chinese-style sausage during refrigerated storage. *Meat Sci.* **2013**, *95*, 145–150. [CrossRef]
17. Saricoban, C.; Özcan, M. Antioxidative activity of Rosemary (*Rosmarinus officinalis* L.) and Sage (*Salvia fruticose* L.) Essential oils in Chicken fat. *J. Essent. Oil Bear. Plants* **2004**, *7*, 91–95. [CrossRef]
18. Šojić, B.; Ikonić, P.; Pavlić, B.; Zeković, Z.; Tomović, V.; Kocić-Tanackov, S.; Džinić, N.; Škaljac, S.; Ivić, M.; Jokanović, M.; et al. The effect of essential oil from sage (*Salvia officinalis* L.) herbal dust (food industry by-product) on the microbiological stability of fresh pork sausages. *IOP Conf. Ser. Earth Environ. Sci.* **2017**, *85*, 12055. [CrossRef]
19. Marangoni, C.; de Moura, N.F. Antioxidant activity of essential oil from *Coriandrum sativum* L. in Italian salami. *Food Sci. Technol.* **2011**, *31*, 124–128. [CrossRef]
20. Michalczyk, M.; Macura, R.; Tesarowicz, I.; Banaś, J. Effect of adding essential oils of coriander (*Coriandrum sativum* L.) and hyssop (*Hyssopus officinalis* L.) on the shelf life of ground beef. *Meat Sci.* **2012**, *90*, 842–850. [CrossRef]
21. Rattanachaikunsopon, P.; Phumkhachorn, P. Potential of Coriander (*Coriandrum sativum*) Oil as a Natural Antimicrobial Compound in Controlling Campylobacter jejuniin Raw Meat. *Biosci. Biotechnol. Biochem.* **2010**, *74*, 31–35. [CrossRef]

22. Shadrack, O.; Konyole, S.; Ngala, S. Effects of Rosemary Spice (*Rosmarinus officinalis* L.) and Nitrite Picking Salt Combination on Keeping and Organoleptic Quality of Beef Sausages. *J. Basic Appl. Sci. Res.* **2012**, *2*, 4008–4015.
23. Nassu, R.T.; Gonçalves, L.A.; Pereira da Silva, M.A.; Beserra, F.J. Oxidative stability of fermented goat meat sausage with different levels of natural antioxidant. *Meat Sci.* **2003**, *63*, 43–49. [CrossRef]
24. Kirkin, C.; Inbat, S.M.; Nikolov, D.; Yildirim, S. Effects of tarragon essential oil on some characteristics of frankfurter type sausages. *AIMS Agric. Food* **2019**, *4*, 244–250. [CrossRef]
25. Mahmoud, S.; Masoud, W. The antibacterial activity of Laurus nobilis leaf extract and its potential use as a preservative for fresh lamb meat. *Afr. J. Microbiol. Res.* **2020**, *14*, 617–624. [CrossRef]
26. Da Silveira, S.M.; Luciano, F.B.; Fronza, N.; Cunha, A.; Scheuermann, G.N.; Vieira, C.R.W. Chemical composition and antibacterial activity of Laurus nobilis essential oil towards foodborne pathogens and its application in fresh Tuscan sausage stored at 7 °C. *LWT Food Sci. Technol.* **2014**, *59*, 86–93. [CrossRef]
27. Jayari, A.; El Abed, N.; Jouini, A.; Abdul-Wahab, O.M.S.; Maaroufi, A.; Ahmed, S.B.H. Antibacterial activity of Thymus capitatus and Thymus algeriensis essential oils against four food-borne pathogens inoculated in minced beef meat. *J. Food Saf.* **2017**, *38*, 12409. [CrossRef]
28. Moarefian, M.; Barzegar, M.; Sattari, M.; Naghdi Badi, H. Production of functional cooked sausage by Mentha piperita essential oil as a natural antioxidant and antimicrobial material. *J. Med. Plants* **2012**, *11*, 46–57.
29. Kunnumakkara, A.B.; Rana, V.; Parama, D.; Banik, K.; Girisa, S.; Henamayee, S.; Thakur, K.K.; Dutta, U.; Garodia, P.; Gupta, S.C.; et al. COVID-19, cytokines, inflammation, and spices: How are they related? *Life Sci.* **2021**, *284*, 119201. [CrossRef]
30. Filipčev, B. Chapter 16—The effects of aromatic plants and their extracts in food products. In *Feed Additives*; Florou-Paneri, P., Christaki, E., Giannenas, I., Eds.; Academic Press: New York, NY, USA, 2020; pp. 279–294. [CrossRef]
31. Kumar, V.; Marković, T.; Emerald, M.; Dey, A. *Herbs: Composition and Dietary Importance*; Academic Press: Oxford, UK, 2015; pp. 332–337. [CrossRef]
32. Agyare, C.; Appiah, T.; Boakye, Y.; Apenteng, J. Petroselinum Crispum: A Review. In *Medicinal Spices and Vegetables from Africa*; Academic Press: New York, NY, USA, 2017; pp. 527–547. [CrossRef]
33. Farzaei, M.H.; Abbasabadi, Z.; Ardekani, M.R.S.; Rahimi, R.; Farzaei, F. Parsley: A review of ethnopharmacology, phytochemistry and biological activities. *J. Tradit. Chin. Med.* **2013**, *33*, 815–826. [CrossRef]
34. Agus, A.; Denizot, J.; Thévenot, J.; Martinez-Medina, M.; Massier, S.; Sauvanet, P.; Bernalier-Donadille, A.; Denis, S.; Hofman, P.; Bonnet, R.; et al. Western diet induces a shift in microbiota composition enhancing susceptibility to Adherent-Invasive *E. coli* infection and intestinal inflammation. *Sci. Rep.* **2016**, *6*, 19032. [CrossRef] [PubMed]
35. Van Praagh, J.B.; Luo, J.N.; Zaborina, O.; Alverdy, J.C. Involvement of the Commensal Organism Bacillus subtilis in the Pathogenesis of Anastomotic Leak. *Surg. Infect.* **2020**, *21*, 865–870. [CrossRef]
36. Zhang, H.; Chen, F.; Wang, X.; Yao, H.-Y. Evaluation of antioxidant activity of parsley (Petroselinum crispum) essential oil and identification of its antioxidant constituents. *Food Res. Int.* **2006**, *39*, 833–839. [CrossRef]
37. Wong, P.Y.Y.; Kitts, D.D. Studies on the dual antioxidant and antibacterial properties of parsley (*Petroselinum crispum*) and cilantro (*Coriandrum sativum*) extracts. *Food Chem.* **2006**, *97*, 505–515. [CrossRef]
38. Vosoughi, N.; Gomarian, M.; Pirbalouti, A.G.; Khaghani, S.; Malekpoor, F. Essential oil composition and total phenolic, flavonoid contents, and antioxidant activity of sage (*Salvia officinalis* L.) extract under chitosan application and irrigation frequencies. *Ind. Crop. Prod.* **2018**, *117*, 366–374. [CrossRef]
39. Borrás-Linares, I.; Stojanović, Z.; Quirantes-Piné, R.; Arráez-Román, D.; Švarc-Gajić, J.; Fernández-Gutiérrez, A.; Segura-Carretero, A. Rosmarinus Officinalis Leaves as a Natural Source of Bioactive Compounds. *Int. J. Mol. Sci.* **2014**, *15*, 20585–20606. [CrossRef]
40. Begum, A.; Sandhya, S.; Ali, S.S.; Vinod, K.R.; Reddy, S.; Banji, D. An in-depth review on the medicinal flora Rosmarinus officinalis (Lamiaceae). *Acta Sci. Pol. Technol. Aliment.* **2014**, *12*, 61–73.
41. De Macedo, L.M.; Santos, É.D.; Militão, L.; Tundisi, L.L.; Ataide, J.A.; Souto, E.B.; Mazzola, P.G. Rosemary (*Rosmarinus officinalis* L.; syn Salvia rosmarinus Spenn.) and Its Topical Applications: A Review. *Plants* **2020**, *9*, 651. [CrossRef]
42. De Oliveira, J.R.; Camargo, S.E.A.; de Oliveira, L.D. *Rosmarinus officinalis* L. (rosemary) as therapeutic and prophylactic agent. *J. Biomed. Sci.* **2019**, *26*, 5. [CrossRef]
43. Cattaneo, L.; Cicconi, R.; Mignogna, G.; Giorgi, A.; Mattei, M.; Graziani, G.; Ferracane, R.; Grosso, A.; Aducci, P.; Schininà, M.E.; et al. Anti-Proliferative Effect of *Rosmarinus officinalis* L. Extract on Human Melanoma A375 Cells. *PLoS ONE* **2015**, *10*, 0132439. [CrossRef]
44. Hernández-Hernández, E.; Lira-Moreno, C.Y.; Guerrero-Legarreta, I.; Wild-Padua, G.; Di Pierro, P.; García-Almendárez, B.E.; Regalado-González, C. Effect of Nanoemulsified and Microencapsulated Mexican Oregano (Lippia graveolens Kunth) Essential Oil Coatings on Quality of Fresh Pork Meat. *J. Food Sci.* **2017**, *82*, 1423–1432. [CrossRef]
45. Krkić, N.; Šojić, B.; Lazić, V.; Petrović, L.; Mandić, A.; Sedej, I.; Tomović, V. Lipid oxidative changes in chitosan-oregano coated traditional dry fermented sausage Petrovská klobása. *Meat Sci.* **2013**, *93*, 767–770. [CrossRef] [PubMed]
46. Doolaege, E.H.; Vossen, E.; Raes, K.; De Meulenaer, B.; Verhé, R.; Paelinck, H.; De Smet, S. Effect of rosemary extract dose on lipid oxidation, colour stability and antioxidant concentrations, in reduced nitrite liver pâtés. *Meat Sci.* **2012**, *90*, 925–931. [CrossRef]
47. Sebranek, J.; Sewalt, V.; Robbins, K.L.; Houser, T.A. Comparison of a natural rosemary extract and BHA/BHT for relative antioxidant effectiveness in pork sausage. *Meat Sci.* **2005**, *69*, 289–296. [CrossRef] [PubMed]

48. Mohamed, H.M.H.; Mansour, H.A. Incorporating essential oils of marjoram and rosemary in the formulation of beef patties manufactured with mechanically deboned poultry meat to improve the lipid stability and sensory attributes. *LWT Food Sci. Technol.* **2012**, *45*, 79–87. [CrossRef]
49. Oswell, N.J.; Thippareddi, H.; Pegg, R.B. Practical use of natural antioxidants in meat products in the U.S.: A review. *Meat Sci.* **2018**, *145*, 469–479. [CrossRef] [PubMed]
50. Babovic, N.; Djilas, S.; Jadranin, M.; Vajs, V.; Ivanovic, J.; Petrovic, S.; Zizovic, I. Supercritical carbon dioxide extraction of antioxidant fractions from selected Lamiaceae herbs and their antioxidant capacity. *Innov. Food Sci. Emerg. Technol.* **2010**, *11*, 98–107. [CrossRef]
51. Berdahl, D.R.; McKeague, J. Rosemary and sage extracts as antioxidants for food preservation. In *Handbook of Antioxidants for Food Preservation*; Elsevier: Oxford, UK, 2015; pp. 177–217.
52. Miraj, S.; Kiani, S. A review study of therapeutic effects of *Salvia officinalis* L. *Pharm. Lett.* **2016**, *8*, 299–303.
53. Ghorbani, A.; Esmaeilizadeh, M. Pharmacological properties of Salvia officinalis and its components. *J. Tradit. Complement. Med.* **2017**, *7*, 433–440. [CrossRef]
54. Pedro, D.F.N.; Ramos, A.A.; Lima, C.F.; Baltazar, F.; Pereira-Wilson, C. Colon Cancer Chemoprevention by Sage Tea Drinking: Decreased DNA Damage and Cell Proliferation. *Phytother. Res.* **2016**, *30*, 298–305. [CrossRef]
55. Miroddi, M.; Navarra, M.; Quattropani, M.C.; Calapai, F.; Gangemi, S.; Calapai, G. Systematic Review of Clinical Trials Assessing Pharmacological Properties ofSalviaSpecies on Memory, Cognitive Impairment and Alzheimer's Disease. *CNS Neurosci. Ther.* **2014**, *20*, 485–495. [CrossRef]
56. Kontogianni, V.G.; Tomic, G.; Nikolić, I.; Nerantzaki, A.A.; Sayyad, N.; Stosic-Grujicic, S.; Stojanović, I.; Gerothanassis, I.P.; Tzakos, A.G. Phytochemical profile of Rosmarinus officinalis and *Salvia officinalis* extracts and correlation to their antioxidant and anti-proliferative activity. *Food Chem.* **2013**, *136*, 120–129. [CrossRef]
57. Kaur, N.; Chahal, K.K.; Kumar, A.; Singh, R.; Bhardwaj, U. Antioxidant activity of *Anethum graveolens* L. essential oil constituents and their chemical analogues. *J. Food Biochem.* **2019**, *43*, 12782. [CrossRef]
58. Hosseinzadeh, H.; Karimi, G.R.; Ameri, M. Effects of Anethum graveolens L. seed extracts on experimental gastric irritation models in mice. *BMC Pharmacol.* **2002**, *2*, 21. [CrossRef]
59. Shekhawat, G.; Jana, S. Anethum graveolens: An Indian traditional medicinal herb and spice. *Pharmacogn. Rev.* **2010**, *4*, 179–184. [CrossRef]
60. Takahashi, N.; Yao, L.; Kim, M.; Sasako, H.; Aoyagi, M.; Shono, J.; Tsuge, N.; Goto, T.; Kawada, T. Dill seed extract improves abnormalities in lipid metabolism through peroxisome proliferator-activated receptor-α (PPAR-α) activation in diabetic obese mice. *Mol. Nutr. Food Res.* **2013**, *57*, 1295–1299. [CrossRef]
61. Al-Sheddi, E.S.; Al-Zaid, N.A.; Al-Oqail, M.M.; Al-Massarani, S.M.; El-Gamal, A.A.; Farshori, N.N. Evaluation of cytotoxicity, cell cycle arrest and apoptosis induced by *Anethum graveolens* L. essential oil in human hepatocellular carcinoma cell line. *Saudi Pharm. J.* **2019**, *27*, 1053–1060. [CrossRef]
62. Orhan, I.E.; Senol, F.S.; Ozturk, N.; Celik, S.A.; Pulur, A.; Kan, Y. Phytochemical contents and enzyme inhibitory and antioxidant properties of *Anethum graveolens* L. (dill) samples cultivated under organic and conventional agricultural conditions. *Food Chem. Toxicol.* **2013**, *59*, 96–103. [CrossRef]
63. Lombrea, A.; Antal, D.; Ardelean, F.; Avram, S.; Pavel, I.Z.; Vlaia, L.; Mut, A.-M.; Diaconeasa, Z.; Dehelean, C.A.; Soica, C.; et al. A Recent Insight Regarding the Phytochemistry and Bioactivity of *Origanum vulgare* L. Essential Oil. *Int. J. Mol. Sci.* **2020**, *21*, 9653. [CrossRef]
64. Morshedloo, M.R.; Craker, L.E.; Salami, A.; Nazeri, V.; Sang, H.; Maggi, F. Effect of prolonged water stress on essential oil content, compositions and gene expression patterns of mono- and sesquiterpene synthesis in two oregano (*Origanum vulgare* L.) subspecies. *Plant Physiol. Biochem.* **2017**, *111*, 119–128. [CrossRef]
65. Han, F.; Ma, G.-Q.; Yang, M.; Yan, L.; Xiong, W.; Shu, J.-C.; Zhao, Z.-D.; Xu, H.-L. Chemical composition and antioxidant activities of essential oils from different parts of the oregano. *J. Zhejiang Univ. Sci. B* **2017**, *18*, 79–84. [CrossRef]
66. Vujicic, M.; Nikolic, I.; Kontogianni, V.G.; Saksida, T.; Charisiadis, P.; Vasic, B.; Stosic-Grujicic, S.; Gerothanassis, I.P.; Tzakos, A.G.; Stojanović, I. Ethyl Acetate Extract of *Origanum vulgare* L. ssp. hirtum Prevents Streptozotocin-Induced Diabetes in C57BL/6 Mice. *J. Food Sci.* **2016**, *81*, 1846–1853. [CrossRef] [PubMed]
67. Veenstra, J.P.; Johnson, J.J. Oregano (Origanium Vulgare) Extract for Food Preservation and Improving Gastrointestinal Health. *Int. J. Nutr.* **2019**, *3*, 43–52. [CrossRef] [PubMed]
68. Balusamy, S.R.; Perumalsamy, H.; Huq, M.A.; Balasubramanian, B. Anti-proliferative activity of Origanum vulgare inhibited lipogenesis and induced mitochondrial mediated apoptosis in human stomach cancer cell lines. *Biomed. Pharmacother.* **2018**, *108*, 1835–1844. [CrossRef] [PubMed]
69. Zhao, Z.; Yin, Z.; Zhao, Q. Red and processed meat consumption and gastric cancer risk: A systematic review and meta-analysis. *Oncotarget* **2017**, *8*, 30563–30575. [CrossRef]
70. Vasconcelos, N.G.; Croda, J.; Silva, K.E.; Motta, M.L.L.; Maciel, W.G.; Limiere, L.C.; Simionatto, S. *Origanum vulgare* L. essential oil inhibits the growth of carbapenem-resistant gram-negative bacteria. *Rev. Soc. Bras. Med. Trop.* **2019**, *52*, 20180502. [CrossRef]
71. Amaral, S.C.; Pruski, B.B.; de Freitas, S.B.; Allend, S.O.; Ferreira, M.R.A.; Moreira, C., Jr.; Pereira, D.I.B.; Junior, A.S.V.; Hartwig, D.D. Origanum vulgare essential oil: Antibacterial activities and synergistic effect with polymyxin B against multidrug-resistant Acinetobacter baumannii. *Mol. Biol. Rep.* **2020**, *47*, 9615–9625. [CrossRef]

72. Dandachi, I.; Chabou, S.; Daoud, Z.; Rolain, J.-M. Prevalence and Emergence of Extended-Spectrum Cephalosporin-, Carbapenem- and Colistin-Resistant Gram Negative Bacteria of Animal Origin in the Mediterranean Basin. *Front. Microbiol.* **2018**, *9*, 2299. [CrossRef]
73. Lupo, A.; Vogt, D.; Seiffert, S.N.; Endimiani, A.; Perreten, V. Antibiotic Resistance and Phylogenetic Characterization of Acinetobacter baumannii Strains Isolated from Commercial Raw Meat in Switzerland. *J. Food Prot.* **2014**, *77*, 1976–1981. [CrossRef]
74. Zheng, W.; Wang, S.Y. Antioxidant Activity and Phenolic Compounds in Selected Herbs. *J. Agric. Food Chem.* **2001**, *49*, 5165–5170. [CrossRef]
75. Fasseas, M.K.; Mountzouris, K.C.; Tarantilis, P.A.; Polissiou, M.; Zervas, G. Antioxidant activity in meat treated with oregano and sage essential oils. *Food Chem.* **2008**, *106*, 1188–1194. [CrossRef]
76. Šimat, V.; Čagalj, M.; Skroza, D.; Gardini, F.; Tabanelli, G.; Montanari, C.; Hassoun, A.; Ozogul, F. Sustainable sources for antioxidant and antimicrobial compounds used in meat and seafood products. *Adv. Food Nutr. Res.* **2021**, *97*, 55–118. [CrossRef]
77. Ghasemzadeh, A.; Ashkani, S.; Baghdadi, A.; Pazoki, A.; Jaafar, H.Z.E.; Rahmat, A. Improvement in Flavonoids and Phenolic Acids Production and Pharmaceutical Quality of Sweet Basil (*Ocimum basilicum* L.) by Ultraviolet-B Irradiation. *Molecules* **2016**, *21*, 1203. [CrossRef]
78. Shirazi, M.T.; Gholami, H.; Kavoosi, G.; Rowshan, V.; Tafsiry, A. Chemical composition, antioxidant, antimicrobial and cytotoxic activities of T agetes minuta and O cimum basilicum essential oils. *Food Sci. Nutr.* **2014**, *2*, 146–155. [CrossRef]
79. Nazir, M.; Tungmunnithum, D.; Bose, S.; Drouet, S.; Garros, L.; Giglioli-Guivarc'h, N.; Abbasi, B.H.; Hano, C. Differential Production of Phenylpropanoid Metabolites in Callus Cultures of *Ocimum basilicum* L. with Distinct In Vitro Antioxidant Activities and In Vivo Protective Effects against UV stress. *J. Agric. Food Chem.* **2019**, *67*, 1847–1859. [CrossRef]
80. Chaudhary, S.; Semwal, A.; Kumar, H.; Verma, H.C.; Kumar, A. In-vivo study for anti-hyperglycemic potential of aqueous extract of Basil seeds (*Ocimum basilicum* Linn) and its influence on biochemical parameters, serum electrolytes and haematological indices. *Biomed. Pharmacother.* **2016**, *84*, 2008–2013. [CrossRef]
81. Al-Subhi, L.; Wa, M.I. Two Cultivars of Ocimum basilicum Leaves Extracts Attenuate Streptozotocin-mediated Oxidative Stress in Diabetic Rats. *Pak. J. Biol. Sci.* **2020**, *23*, 1010–1017. [CrossRef]
82. Almalki, D.A. Renoprotective Effect of Ocimum Basilicum (Basil) Against Diabetes-induced Renal Affection in Albino Rats. *Mater. Socio-Med.* **2019**, *31*, 236–240. [CrossRef]
83. Teofilović, B.; Grujić-Letić, N.; Goločorbin-Kon, S.; Stojanović, S.; Vastag, G.; Gadžurić, S. Experimental and chemometric study of antioxidant capacity of basil (*Ocimum basilicum*) extracts. *Ind. Crop. Prod.* **2017**, *100*, 176–182. [CrossRef]
84. Chaves, R.D.S.B.; Martins, R.L.; Rodrigues, A.B.L.; de Menezes Rabelo, É.; Farias, A.L.F.; Brandão, L.B.; Santos, L.L.; Galardo, A.K.R.; de Almeida, S.S.M.d. Evaluation of larvicidal potential against larvae of Aedes aegypti (Linnaeus, 1762) and of the antimicrobial activity of essential oil obtained from the leaves of *Origanum majorana* L. *PLoS ONE* **2020**, *15*, 0235740. [CrossRef]
85. Bina, F.; Rahimi, R. Sweet Marjoram: A Review of Ethnopharmacology, Phytochemistry, and Biological Activities. *J. Evid. Based. Complement. Altern. Med.* **2017**, *22*, 175–185. [CrossRef]
86. Badran, A.; Baydoun, E.; Samaha, A.; Pintus, G.; Mesmar, J.; Iratni, R.; Issa, K.; Eid, A.H. Marjoram Relaxes Rat Thoracic Aorta Via a PI3-K/eNOS/cGMP Pathway. *Biomolecules* **2019**, *9*, 227. [CrossRef]
87. Duletic, S.; Alimpić Aradski, A.; Kolarevic, S.; Vuković-Gačić, B.; Oaldje, M.; Živković, J.; Šavikin, K.; Marin, P. Antineurodegenerative, antioxidant and antibacterial activities and phenolic components of *Origanum majorana* L. (Lamiaceae) extracts. *J. Appl. Bot. Food Qual.* **2018**, *91*, 126–134. [CrossRef]
88. Bouyahya, A.; Chamkhi, I.; Benali, T.; Guaouguaou, F.-E.; Balahbib, A.; El Omari, N.; Taha, D.; Belmehdi, O.; Ghokhan, Z.; El Menyiy, N. Traditional use, phytochemistry, toxicology, and pharmacology of *Origanum majorana* L. *J. Ethnopharmacol.* **2020**, *265*, 113318. [CrossRef]
89. Thanh, V.M.; Bui, L.M.; Bach, L.G.; Nguyen, N.T.; Le Thi, H.; Thi, T.T.H. *Origanum majorana* L. Essential Oil-Associated Polymeric Nano Dendrimer for Antifungal Activity against Phytophthora infestans. *Materials* **2019**, *12*, 1446. [CrossRef]
90. Gomes, F.; Dias, M.I.; Lima, Â.; Barros, L.; Rodrigues, M.E.; Ferreira, I.C.F.R.; Henriques, M. *Satureja montana* L. and *Origanum majorana* L. Decoctions: Antimicrobial Activity, Mode of Action and Phenolic Characterization. *Antibiotics* **2020**, *9*, 294. [CrossRef]
91. Schmidt, E.; Bail, S.; Buchbauer, G.; Stoilova, I.; Krastanov, A.; Stoyanova, A.; Jirovetz, L. Chemical Composition, Olfactory Evaluation and Antioxidant Effects of the Essential oil of Origanum Majorana L. from Albania. *Nat. Prod. Commun.* **2008**, *3*, 1051–1056. [CrossRef]
92. Beigi, M.; Torki-Harchegani, M.; Pirbalouti, A.G. Quantity and chemical composition of essential oil of peppermint (*Mentha* × *piperita* L.) leaves under different drying methods. *Int. J. Food Prop.* **2018**, *21*, 267–276. [CrossRef]
93. Bassolé, I.H.N.; Juliani, H.R. Essential Oils in Combination and Their Antimicrobial Properties. *Molecules* **2012**, *17*, 3989–4006. [CrossRef]
94. Mahendran, G.; Rahman, L.U. Ethnomedicinal, phytochemical and pharmacological updates on Peppermint (*Mentha* × *piperita* L.)—A review. *Phytother. Res.* **2020**, *34*, 2088–2139. [CrossRef]
95. Kennedy, D.; Okello, E.; Chazot, P.; Howes, M.-J.; Ohiomokhare, S.; Jackson, P.; Haskell-Ramsay, C.; Khan, J.; Forster, J.; Wightman, E. Volatile Terpenes and Brain Function: Investigation of the Cognitive and Mood Effects of *Mentha* × *piperita* L. Essential Oil with In Vitro Properties Relevant to Central Nervous System Function. *Nutrients* **2018**, *10*, 1029. [CrossRef]
96. Kanatt, S.R.; Chander, R.; Sharma, A. Antioxidant potential of mint (*Mentha spicata* L.) in radiation-processed lamb meat. *Food Chem.* **2007**, *100*, 451–458. [CrossRef]

97. Kanatt, S.R.; Chander, R.; Sharma, A. Chitosan and mint mixture: A new preservative for meat and meat products. *Food Chem.* **2008**, *107*, 845–852. [CrossRef]
98. Biswas, A.K.; Chatli, M.K.; Sahoo, J. Antioxidant potential of curry (*Murraya koenigii* L.) and mint (Mentha spicata) leaf extracts and their effect on colour and oxidative stability of raw ground pork meat during refrigeration storage. *Food Chem.* **2012**, *133*, 467–472. [CrossRef] [PubMed]
99. Zawiślak, G.; Dzida, K. Composition of essential oils and content of macronutrients in herbage of tarragon (*Artemisia dracunculus* L.) grown in south-eastern Poland. *J. Elem.* **2012**, *17*, 721–729. [CrossRef]
100. Duric, K.; Kovac-Besovic, E.E.; Niksic, H.; Muratović, S.; Sofic, E. Anticoagulant activity of some Artemisia dracunculus leaf extracts. *Bosn. J. Basic Med. Sci.* **2015**, *15*, 9–14. [CrossRef]
101. Vandanmagsar, B.; Haynie, K.R.; Wicks, S.E.; Bermudez, E.M.; Mendoza, T.M.; Ribnicky, D.; Cefalu, W.T.; Mynatt, R.L. *Artemisia dracunculus* L. extract ameliorates insulin sensitivity by attenuating inflammatory signalling in human skeletal muscle culture. *Diabetes Obes. Metab.* **2014**, *16*, 728–738. [CrossRef]
102. Behbahani, B.A.; Shahidi, F.; Yazdi, F.T.; Mortazavi, S.A.; Mohebbi, M. Antioxidant activity and antimicrobial effect of tarragon (*Artemisia dracunculus*) extract and chemical composition of its essential oil. *J. Food Meas. Charact.* **2017**, *11*, 847–863. [CrossRef]
103. Nimse, S.B.; Pal, D. Free radicals, natural antioxidants, and their reaction mechanisms. *RSC Adv.* **2015**, *5*, 27986–28006. [CrossRef]
104. Bhat, S.; Kaushal, P.; Kaur, M.; Sharma, H.K. Coriander (*Coriandrum sativum* L.): Processing, nutritional and functional aspects. *Afr. J. Plant Sci.* **2014**, *8*, 25–33. [CrossRef]
105. Al-Mofleh, I.A.; Alhaider, A.A.; Mossa, J.S.; Al-Sohaibani, M.O.; Rafatullah, S.; Qureshi, S. Protection of gastric mucosal damage by *Coriandrum sativum* L. pretreatment in Wistar albino rats. *Environ. Toxicol. Pharmacol.* **2006**, *22*, 64–69. [CrossRef]
106. Freires, I.; Murata, R.M.; Furletti, V.F.; Sartoratto, A.; de Alencar, S.M.; Figueira, G.M.; Rodrigues, J.A.D.O.; Duarte, M.C.T.; Rosalen, P.L. *Coriandrum sativum* L. (Coriander) Essential Oil: Antifungal Activity and Mode of Action on *Candida* spp., and Molecular Targets Affected in Human Whole-Genome Expression. *PLoS ONE* **2014**, *9*, 99086. [CrossRef]
107. Rodrigues, K.E.; de Oliveira, F.R.; Barbosa, B.R.C.; Paraense, R.S.O.; Bannwart, C.M.; Pinheiro, B.G.; Botelho, A.D.S.; Muto, N.A.; Amarante, C.B.D.; Hamoy, M.; et al. Aqueous *Coriandrum sativum* L. extract promotes neuroprotection against motor changes and oxidative damage in rat progeny after maternal exposure to methylmercury. *Food Chem. Toxicol.* **2019**, *133*, 110755. [CrossRef]
108. Mani, V.; Parle, M.; Ramasamy, K.; Majeed, A.B.A. Reversal of memory deficits by Coriandrum sativum leaves in mice. *J. Sci. Food Agric.* **2011**, *91*, 186–192. [CrossRef]
109. Aissaoui, A.; Zizi, S.; Israili, Z.H.; Lyoussi, B. Hypoglycemic and hypolipidemic effects of *Coriandrum sativum* L. in Meriones shawi rats. *J. Ethnopharmacol.* **2011**, *137*, 652–661. [CrossRef]
110. Šojić, B.; Pavlić, B.; Ikonić, P.; Tomović, V.; Ikonić, B.; Zeković, Z.; Kocić-Tanackov, S.; Jokanović, M.; Škaljac, S.; Ivić, M. Coriander essential oil as natural food additive improves quality and safety of cooked pork sausages with different nitrite levels. *Meat Sci.* **2019**, *157*, 107879. [CrossRef]
111. Pavlić, B.; Vidović, S.; Vladić, J.; Radosavljević, R.; Zeković, Z. Isolation of coriander (*Coriandrum sativum* L.) essential oil by green extractions versus traditional techniques. *J. Supercrit. Fluids* **2015**, *99*, 23–28. [CrossRef]
112. Řebíčková, K.; Bajer, T.; Šilha, D.; Ventura, K.; Bajerová, P. Comparison of Chemical Composition and Biological Properties of Essential Oils Obtained by Hydrodistillation and Steam Distillation of *Laurus nobilis* L. *Mater. Veg.* **2020**, *75*, 495–504. [CrossRef]
113. Peixoto, L.R.; Rosalen, P.L.; Ferreira, G.L.S.; Freires, I.A.; de Carvalho, F.G.; Castellano, L.R.; de Castro, R.D. Antifungal activity, mode of action and anti-biofilm effects of Laurus nobilis Linnaeus essential oil against Candida spp. *Arch. Oral Biol.* **2017**, *73*, 179–185. [CrossRef]
114. Speroni, E.; Cervellati, R.; Dall'Acqua, S.; Guerra, M.C.; Greco, E.; Govoni, P.; Innocenti, G. Gastroprotective Effect and Antioxidant Properties of DifferentLaurus nobilisL. Leaf Extracts. *J. Med. Food* **2011**, *14*, 499–504. [CrossRef]
115. Ramos, C.; Teixeira, B.; Batista, I.; Matos, O.; Serrano, M.D.C.; Neng, N.; Nogueira, J.; Nunes, M.L.; Marques, A. Antioxidant and antibacterial activity of essential oil and extracts of bay laurel Laurus nobilis Linnaeus (Lauraceae) from Portugal. *Nat. Prod. Res.* **2012**, *26*, 518–529. [CrossRef]
116. Djenane, D.; Yangüela, J.; Gómez, D.; Roncalés, P. Perspectives on the use of essential oils as antimicrobials against campylobacter jejuni cect 7572 in retail chicken meats packaged in microaerobic atmosphere. *J. Food Saf.* **2012**, *32*, 37–47. [CrossRef]
117. Nieto, G. A Review on Applications and Uses of *Thymus* in the Food Industry. *Plants* **2020**, *9*, 961. [CrossRef]
118. Fachini-Queiroz, F.C.; Kummer, R.; Estevão-Silva, C.F.; Carvalho, M.D.D.B.; Cunha, J.M.; Grespan, R.; Bersani-Amado, C.A.; Cuman, R.K.N. Effects of Thymol and Carvacrol, Constituents of *Thymus vulgaris* L. Essential Oil, on the Inflammatory Response. *Evid. Based Complement. Altern. Med.* **2012**, *2012*, 657026. [CrossRef]
119. Roby, M.H.H.; Sarhan, M.A.; Selim, K.A.-H.; Khalel, K.I. Evaluation of antioxidant activity, total phenols and phenolic compounds in thyme (*Thymus vulgaris* L.), sage (*Salvia officinalis* L.), and marjoram (*Origanum majorana* L.) extracts. *Ind. Crop. Prod.* **2013**, *43*, 827–831. [CrossRef]
120. Oliviero, M.; Romilde, I.; Beatrice, M.M.; Matteo, V.; Giovanna, N.; Consuelo, A.; Claudio, C.; Giorgio, S.; Filippo, M.; Massimo, N. Evaluations of thyme extract effects in human normal bronchial and tracheal epithelial cell lines and in human lung cancer cell line. *Chem. Interact.* **2016**, *256*, 125–133. [CrossRef]
121. Asadbegi, M.; Yaghmaei, P.; Salehi, I.; Komaki, A.; Ebrahim-Habibi, A. Investigation of thymol effect on learning and memory impairment induced by intrahippocampal injection of amyloid beta peptide in high fat diet- fed rats. *Metab. Brain Dis.* **2017**, *32*, 827–839. [CrossRef]

122. Serrano, A.; González-Sarrías, A.; Tomás-Barberán, F.A.; Avellaneda, A.; Gironés-Vilaplana, A.; Nieto, G.; Ros-Berruezo, G. Anti-Inflammatory and Antioxidant Effects of Regular Consumption of Cooked Ham Enriched with Dietary Phenolics in Diet-Induced Obese Mice. *Antioxidants* **2020**, *9*, 639. [CrossRef]
123. Soković, M.; Glamočlija, J.; Marin, P.D.; Brkić, D.; Van Griensven, L.J.L.D. Antibacterial Effects of the Essential Oils of Commonly Consumed Medicinal Herbs Using an In Vitro Model. *Molecules* **2010**, *15*, 7532–7546. [CrossRef]
124. Aureli, P.; Costantini, A.; Zolea, S. Antimicrobial Activity of Some Plant Essential Oils against Listeria monocytogenes. *J. Food Prot.* **1992**, *55*, 344–348. [CrossRef]
125. Fratianni, F.; De Martino, L.; Melone, A.; De Feo, V.; Coppola, R.; Nazzaro, F. Preservation of Chicken Breast Meat Treated with Thyme and Balm Essential Oils. *J. Food Sci.* **2010**, *75*, M528–M535. [CrossRef]
126. Zengin, H.; Baysal, A.H. Antioxidant and Antimicrobial Activities of Thyme and Clove Essential Oils and Application in Minced Beef. *J. Food Process. Preserv.* **2015**, *39*, 1261–1271. [CrossRef]
127. Dhifi, W.; Bellili, S.; Jazi, S.; Bahloul, N.; Mnif, W. Essential Oils' Chemical Characterization and Investigation of Some Biological Activities: A Critical Review. *Medicines* **2016**, *3*, 25. [CrossRef]
128. Alirezalu, K.; Pateiro, M.; Yaghoubi, M.; Alirezalu, A.; Peighambardoust, S.H.; Lorenzo, J.M. Phytochemical constituents, advanced extraction technologies and techno-functional properties of selected Mediterranean plants for use in meat products. A comprehensive review. *Trends Food Sci. Technol.* **2020**, *100*, 292–306. [CrossRef]
129. Burt, S. Essential oils: Their antibacterial properties and potential applications in foods—A review. *Int. J. Food Microbiol.* **2004**, *94*, 223–253. [CrossRef] [PubMed]
130. Falleh, H.; Ben Jemaa, M.; Saada, M.; Ksouri, R. Essential oils: A promising eco-friendly food preservative. *Food Chem.* **2020**, *330*, 127268. [CrossRef]
131. Liu, D.-C.; Tsau, R.-T.; Lin, Y.-C.; Jan, S.-S.; Tan, F.-J. Effect of various levels of rosemary or Chinese mahogany on the quality of fresh chicken sausage during refrigerated storage. *Food Chem.* **2009**, *117*, 106–113. [CrossRef]
132. Tongnuanchan, P.; Benjakul, S. Essential Oils: Extraction, Bioactivities, and Their Uses for Food Preservation. *J. Food Sci.* **2014**, *79*, 1231–1249. [CrossRef]
133. Zhang, H.; Wu, J.; Guo, X. Effects of antimicrobial and antioxidant activities of spice extracts on raw chicken meat quality. *Food Sci. Hum. Wellness* **2016**, *5*, 39–48. [CrossRef]
134. Zaouali, Y.; Messaoud, C.; Ben Salah, A.; Boussaïd, M. Oil composition variability among populations in relationship with their ecological areas in Tunisian *Rosmarinus officinalis* L. *Flavour Fragr. J.* **2005**, *20*, 512–520. [CrossRef]
135. Mata, A.T.; Proença, C.; Ferreira, A.R.; Serralheiro, M.L.M.; Nogueira, J.M.F.; Araújo, M.E.M. Antioxidant and antiacetyl-cholinesterase activities of five plants used as Portuguese food spices. *Food Chem.* **2007**, *103*, 778–786. [CrossRef]
136. Porte, A.; Godoy, R.L.D.O.; Lopes, D.; Koketsu, M.; Gonçalves, S.L.; Torquilho, H.S. Essential Oil of *Rosmarinus officinalis* L. (Rosemary) from Rio de Janeiro, Brazil. *J. Essent. Oil Res.* **2000**, *12*, 577–580. [CrossRef]
137. Satyal, P.; Jones, T.H.; Lopez, E.M.; McFeeters, R.L.; Ali, N.A.A.; Mansi, I.; Al-Kaf, A.G.; Setzer, W.N. Chemotypic Characterization and Biological Activity of Rosmarinus officinalis. *Foods* **2017**, *6*, 20. [CrossRef]
138. Serrano, E.; Palma, J.; Tinoco, T.; Venâncio, F.; Martins, A. Evaluation of the Essential Oils of Rosemary (*Rosmarinus officinalis* L.) from Different Zones of "Alentejo" (Portugal). *J. Essent. Oil Res.* **2002**, *14*, 87–92. [CrossRef]
139. Zermane, A.; Meniai, A.-H.; Barth, D. Supercritical CO_2 Extraction of Essential Oil from Algerian Rosemary (*Rosmarinus officinalis* L.). *Chem. Eng. Technol.* **2010**, *33*, 489–498. [CrossRef]
140. Özcan, M.M.; Chalchat, J.-C. Chemical composition and antifungal activity of rosemary (*Rosmarinus officinalis* L.) oil from Turkey. *Int. J. Food Sci. Nutr.* **2008**, *59*, 691–698. [CrossRef]
141. Pitarokili, D.; Tzakou, O.; Loukis, A. Composition of the Essential Oil of Spontaneous Rosmarinus officinalisfrom Greece and Antifungal Activity Against Phytopathogenic Fungi. *J. Essent. Oil Res.* **2008**, *20*, 457–459. [CrossRef]
142. Schmidt, E.; Bail, S.; Buchbauer, G.; Stoilova, I.; Atanasova, T.; Stoyanova, A.; Krastanov, A.; Jirovetz, L. Chemical Composition, Olfactory Evaluation and Antioxidant Effects of Essential Oil from Mentha x piperita. *Nat. Prod. Commun.* **2009**, *4*, 1107–1112. [CrossRef]
143. Khorasaninejad, S. The Effect of Salinity Stress on Growth Parameters, Essential oil Yield and Constituent of Peppermint (*Mentha piperita* L.). *World Appl. Sci. J.* **2010**, *11*, 1403–1407.
144. Horváth, G.; Szabó, L.G.; Lemberkovics, E.; Botz, L.; Kocsis, B. Characterization and TLC-bioautographic detection of essential oils from some Thymustaxa. Determination of the activity of the oils and their components against plant pathogenic bacteria. *JPC—J. Planar Chromatogr. Mod. TLC* **2004**, *17*, 300–304. [CrossRef]
145. Wei, J.-N.; Liu, Z.-H.; Zhao, Y.-P.; Zhao, L.-L.; Xue, T.-K.; Lan, Q.-K. Phytochemical and bioactive profile of *Coriandrum sativum* L. *Food Chem.* **2019**, *286*, 260–267. [CrossRef]
146. Kohara, K.; Sakamoto, Y.; Hasegawa, H.; Kozuka, H.; Sakamoto, K.; Hayata, Y. Fluctuations in Volatile Compounds in Leaves, Stems, and Fruits of Growing Coriander (*Coriandrum sativum* L.) Plants. *J. Jpn. Soc. Hortic. Sci.* **2006**, *75*, 267–269. [CrossRef]
147. Farsaraei, S.; Moghaddam, M.; Pirbalouti, A.G. Changes in growth and essential oil composition of sweet basil in response of salinity stress and superabsorbents application. *Sci. Hortic.* **2020**, *271*, 109465. [CrossRef]
148. Filho, L.C.C.; Martinazzo, A.P.; Teodoro, C.E.D.S.; Vivès, L. Microbiological quality and essential oil of parsley (Petroselinum crispum) submitted to the hygienizing and drying process. *Ind. Crop. Prod.* **2018**, *114*, 180–184. [CrossRef]

149. Sobieszczańska, N.; Myszka, K.; Szwengiel, A.; Majcher, M.; Grygier, A.; Wolko, Ł. Tarragon essential oil as a source of bioactive compounds with anti-quorum sensing and anti-proteolytic activity against *Pseudomonas* spp. isolated from fish—in vitro, in silico and in situ approaches. *Int. J. Food Microbiol.* **2020**, *331*, 108732. [CrossRef]
150. Teixeira, B.; Marques, A.; Ramos, C.; Neng, N.R.; Nogueira, J.M.; Saraiva, J.A.; Nunes, M.L. Chemical composition and antibacterial and antioxidant properties of commercial essential oils. *Ind. Crop. Prod.* **2013**, *43*, 587–595. [CrossRef]
151. Daferera, D.J.; Ziogas, B.N.; Polissiou, M.G. GC-MS Analysis of Essential Oils from Some Greek Aromatic Plants and Their Fungitoxicity on Penicillium digitatum. *J. Agric. Food Chem.* **2000**, *48*, 2576–2581. [CrossRef]
152. Bartoňková, I.; Dvořák, Z. Essential oils of culinary herbs and spices display agonist and antagonist activities at human aryl hydrocarbon receptor AhR. *Food Chem. Toxicol. Int. J. Publ. Br. Ind. Biol. Res. Assoc.* **2018**, *111*, 374–384. [CrossRef] [PubMed]
153. Delaquis, P.J.; Stanich, K.; Girard, B.; Mazza, G. Antimicrobial activity of individual and mixed fractions of dill, cilantro, coriander and eucalyptus essential oils. *Int. J. Food Microbiol.* **2002**, *74*, 101–109. [CrossRef]
154. Martínez, A.L.; González-Trujano, M.E.; Pellicer, F.; López-Muñoz, F.J.; Navarrete, A. Antinociceptive Effect and GC/MS Analysis of *Rosmarinus officinalis* L. Essential Oil from its Aerial Parts. *Planta Med.* **2009**, *75*, 508–511. [CrossRef] [PubMed]
155. Elamrani, A.; Zrira, S.; Benjilali, B.; Berrada, M. A Study of Moroccan Rosemary Oils. *J. Essent. Oil Res.* **2000**, *12*, 487–495. [CrossRef]
156. El-Massry, K.; Farouk, A.; Abou-Zeid, M. Free Radical Scavenging Activity and Lipoxygenase Inhibition of Rosemary (*Rosmarinus officinalis* L.) Volatile Oil. *J. Essent. Oil Bear. Plants* **2008**, *11*, 536–543. [CrossRef]
157. Diab, Y.; Auezova, L.; Chebib, H.; Chalchat, J.-C.; Figueredo, G. Chemical Composition of Lebanese Rosemary (*Rosmarinus officinalis* L.) Essential Oil as a Function of the Geographical Region and the Harvest Time. *J. Essent. Oil Res.* **2002**, *14*, 449–452. [CrossRef]
158. Angioni, A.; Barra, A.; Cereti, E.; Barile, D.; Coïsson, J.D.; Arlorio, M.; Dessi, S.; Coroneo, V.; Cabras, P. Chemical Composition, Plant Genetic Differences, Antimicrobial and Antifungal Activity Investigation of the Essential Oil of *Rosmarinus officinalis* L. *J. Agric. Food Chem.* **2004**, *52*, 3530–3535. [CrossRef]
159. Bouyahya, A.; Lagrouh, F.; El Omari, N.; Bourais, I.; El Jemli, M.; Marmouzi, I.; Salhi, N.; Faouzi, M.E.A.; Belmehdi, O.; Dakka, N.; et al. Essential oils of Mentha viridis rich phenolic compounds show important antioxidant, antidiabetic, dermatoprotective, antidermatophyte and antibacterial properties. *Biocatal. Agric. Biotechnol.* **2020**, *23*, 101471. [CrossRef]
160. Punetha, D.; Tewari, G.; Pande, C. Compositional variability in inflorescence essential oil of Coriandrum sativum from North India. *J. Essent. Oil Res.* **2017**, *30*, 113–119. [CrossRef]
161. Sriti, J.; Talou, T.; Wannes, W.A.; Cerny, M.; Marzouk, B. Essential oil, fatty acid and sterol composition of Tunisian coriander fruit different parts. *J. Sci. Food Agric.* **2009**, *89*, 1659–1664. [CrossRef]
162. Chung, I.-M.; Ahmad, A.; Kim, S.-J.; Naik, P.M.; Nagella, P. Composition of the essential oil constituents from leaves and stems of Korean Coriandrum sativum and their immunotoxicity activity on the *Aedes aegypti* L. *Immunopharmacol. Immunotoxicol.* **2011**, *34*, 152–156. [CrossRef]
163. Marino, M.; Bersani, C.; Comi, G. Impedance measurements to study the antimicrobial activity of essential oils from Lamiaceae and Compositae. *Int. J. Food Microbiol.* **2001**, *67*, 187–195. [CrossRef]

plants

Article

Cytotoxicity Evaluation and Antioxidant Activity of a Novel Drink Based on Roasted Avocado Seed Powder

Andreea Pușcaș [1], Anda E. Tanislav [1], Romina A. Marc [1], Vlad Mureșan [1], Andruța E. Mureșan [1,*], Emoke Pall [2] and Constantin Cerbu [2]

1 Food Engineering Department, Faculty of Food Science and Technology, University of Agricultural Sciences and Veterinary Medicine Cluj-Napoca, 3-5 Calea Mănăştur Street, 400372 Cluj-Napoca, Romania; andreea.puscas@usamvcluj.ro (A.P.); anda.tanislav@usamvcluj.ro (A.E.T.); romina.vlaic@usamvcluj.ro (R.A.M.); vlad.muresan@usamvcluj.ro (V.M.)

2 Department of Infectious Diseases, Faculty of Veterinary Medicine, University of Agricultural Sciences and Veterinary Medicine Cluj-Napoca, 3-5 Calea Mănăştur Street, 400372 Cluj-Napoca, Romania; emoke.pall@usamvcluj.ro (E.P.); constantin.cerbu@usamvcluj.ro (C.C.)

* Correspondence: andruta.muresan@usamvcluj.ro; Tel.: +4-(074)-447-6052

Abstract: The avocado seed is an underused waste resulting from the processing of pulp. Polyphenols, fibers, and carotenoids are present in the seed, which also exhibits prophylactic, fungicidal, and larvicidal effects. Developing food products with avocado seed as an ingredient or spice is highly desired for nutritional, environmental, and economic reasons. The present study proposed its valorization in a hot drink, similar to already existing coffee alternatives, obtained by infusing the roasted and grinded avocado seed. The proximate composition of the raw or conditioned avocado seed and that of the novel drink were determined. The total phenolic content was assessed using the Folin-Ciocâlteu method. The total carotenoids were extracted and assessed spectrophotometrically. Starch determination was performed by the Ewers Polarimetric method. The highest content of polyphenols, 772.90 mg GAE/100 g, was determined in the crude seed, while in the drink was as low as 17.55 mg GAE/100 g. However, the proposed drink demonstrated high antioxidant capacity, evaluated through the DPPH method. This might be due to the high content of the total carotenoid compounds determined in the roasted seed (6534.48 μg/100 g). The proposed drink demonstrated high antiproliferative activity on Hs27 and DLD-1 cell lines.

Keywords: avocado seed; valorization; coffee alternative; food waste; bioactive compounds; antioxidant capacity; cytotoxicity

Citation: Pușcaș, A.; Tanislav, A.E.; Marc, R.A.; Mureșan, V.; Mureșan, A.E.; Pall, E.; Cerbu, C. Cytotoxicity Evaluation and Antioxidant Activity of a Novel Drink Based on Roasted Avocado Seed Powder. *Plants* **2022**, *11*, 1083. https://doi.org/10.3390/plants11081083

Academic Editor: Petko Denev

Received: 8 March 2022
Accepted: 12 April 2022
Published: 15 April 2022

1. Introduction

The peel and seed of the avocado, resulting as by-products in the processing of the pulp, are waste materials which should have application in the domains of food, nutrition and medicine, since recent studies revealed and characterized the valuable nutrients they contain, demonstrating health promoting effects of some of their extracts [1–4]. The peel is edible, but is not consumed due to its bitter taste, chewiness, and because it is hard to digest. The seed is tough, has an astringent taste, and needs processing prior to consumption; therefore, formulating food products containing these valuable ingredients is challenging.

The pulp of avocado is a great source of proteins and monounsaturated fatty acids (predominantly oleic acid—62.14%, palmitic 17.2%, linoleic 11.11%, and palmitoleic acid 7.34%) and low amounts of stearic acid, 0.63%, protecting consumers against coronary heart disease development [5,6]. Besides fatty acids, the nutritional value of avocado is also due to antioxidants, carotenoids, phytochemicals such as α-tocopherol and β-sitosterol, and vitamins (B6, biotin, folic acid, thiamine, riboflavin, vitamin D and K) [7,8]. Avocado pulp has been used (fresh or dehydrated, defatted) in supplementing food products such as meat alternatives or in replacing wheat flour and butter, in whole grain crackers [9,10].

The processed products of avocado pulp include oil, the paste, puree, and guacamole; each of them is sensible to oxidation processes, so preservation or physico-chemical treatments are mandatory. It was reported that increased consumption levels can reduce adult weight gain [11], so it is no surprise avocado is becoming more and more consumed. Along with this, a high level of waste material is also generated on the industrial scale or in individual households.

The seed represents 13–18% of the weight of the whole fruit and the residues have a significant environmental impact due to the great organic charge they contain. Additionally, avocado seeds generate costs associated with disposing, handling, transport, and storage [7,12,13]. Numerous studies revealed some valuable compounds in the chemical composition of the avocado seed, imparting antioxidant and antimicrobial properties to it, transforming this waste material into a valuable ingredient of food products with the potential for medicinal use [3,14,15]. These studies were conducted on various species of avocado, such as Corillo [16] or Mill [17], including the Hass variety [18]. Phytosterols, triterpenes, fatty acids, furanoic acid, abscisic acid, proanthocyanidines (PACs), and other polyphenols are present in variable amounts in the seed, depending on the maturity, growth condition, and variety of the avocado [15,19,20].

Besides, anti-inflammatory, hypoglycemic, antihypertensive, fungicidal, larvicidal, hypolipidemic, analgesic, amoebicidal, and giardicidal activities of some extracts from the avocado seed have been reported [15,21–25]. The seed contains more soluble fibers than the pulp, so by ingesting it, one could naturally prevent constipation. It is also proved to be effective in the prophylaxis of gastric ulcers, preventing the occurrence of this disease [26,27]. Avocado seed is also rich in tannins, carotenoids, and tocopherols, which inhibited the in vitro growth of prostate cancer cell lines [28].

Extracts from the peel and seed displayed different functionalities when added to food products [17]. Acetone/water (70:10 *v/v*) extracts from the peel and seed were included in raw porcine patties and hindered the oxidative reactions and color deterioration during chilled storage of the product [29]. A natural orange colorant was extracted with water from the avocado seed which was priorly grinded [30].

Due to some antinutritional compounds present in the avocado seed, namely phytate, oxalate and cyanogenic glycosides [31], studies have been carried out in order to evaluate the effect of processing methods against the antinutritional compounds of *Persea americana* seed as a step towards establishing purposeful utilization in the food production area [13,32,33]. Solid-state fermentation of Hass avocado seed with *A. niger* GH1 led to an improved antioxidant activity [4]. Among other antinutritional factors, tannins, phytic acid and alkaloids were determined [1]. Soaking and boiling of the avocado seeds reduced the antinutritional compounds to a large extent. Some research describing novel food products with avocado seed as the ingredient has already been conducted [9,33–35]. Powders obtained from the seeds of avocado and peeled ginger roots were used for the preparation of eight prototypes of candies with sugar or aspartame. The candies were analyzed in terms of moisture content (77.3–92.5%), total sugars (0.10–0.66 mg/100 g) and microbial count. The sensorial analysis were promising, the candies being scored above average in taste, texture, and flavor [36].

Avocado seeds have been proposed to be valorized as flour and it would be of special use in tropical countries, were crops such as wheat, barley, millet, and rye do not lead to quality flours and lead to higher costs for pastry products [37]. For the flour preparation, the washed seeds were chopped and dried in an oven at 60 °C for 5 h. The seeds were then grounded in a pulverizer. The avocado seed flour was characterized in terms of yield, proximate analysis, gluten, and falling number and different biscuit formulations were prepared in order to test its applicability. The anthocyanin pigments of the avocado seed led to a darker color for the biscuits, which had an average acceptability in terms of the organoleptic properties. In conclusion, the avocado seed flour, despite the high nutritional value, would not be suitable for baking, having high values for the falling number and not forming the gluten network needed for structure [37]. However, Hass Avocado seed flour alone or in mixture with corn was studied for obtaining extruded snacks, showing promising results [35].

In the current study, drying and roasting were explored as conditioning treatments aiming to reduce the antinutritional compounds activity and concentration in the avocado seed. The present study proposed the use of the avocado seed powder in a hot drink, which could be consumed as a coffee alternative. Novel usages would increase its consumption and decrease food waste and pollution. The roasted avocado seed powder and thereof novel drink were analyzed, and the results revealed some antioxidant functionality and anti-carcinogenic effects.

2. Results

2.1. *Optimization of Roasting of Avocado Seed and the Drink Preparation*

Since the organoleptic properties of avocado seed were unknown, the temperature setting for the heat treatment was started at 135 °C, with a duration varying from 5–90 min. The best results in term of color and flavor were obtained after 90 min, when the seed was dark brown and well flavored. However, in order to reduce the duration of the roasting process, higher-temperature protocols were also explored. Table 1 presents the organoleptic profile of the avocado seed roasted under different time-temperatures protocols, starting from 160 °C ± 5 °C to 200 °C ± 5 °C. Given the results, the most suitable time-temperature protocol was: 180 °C/25 min, respectively 180 °C/30 min and 200 °C/5 min.

Table 1. Organoleptic profile of roasted avocado seed at different temperature-time intervals.

Time [min]	Temperature (°C)					
	160 °C ± 5 °C		180 °C ± 5 °C		200 °C	
	Color	Flavor	Color	Flavor	Color	Flavor
5	unchanged		unchanged	unchanged	dark brown	very flavored
10	light orange	no flavor		slight flavor	dark brown	carbonized
15			light brown	slight flavor	black	
20	dark orange	no flavor	slight dark brown	flavored		
30	light brown	slight flavor	dark brown	well flavored		
35		flavored	very dark brown	carbonized		
45	slight dark brown					
60	dark brown	well flavored				
75	very dark brown	strong flavor				
90	very dark brown	carbonized				

In order to obtain a product with optimal properties, different percentages of water and avocado powder were studied. The final composition was decided to be composed of 93% water and 7% roasted avocado seed powder. The powder had an intense and slightly astringent aroma, and it is sufficient even in this low percentage for obtaining the novel hot drink.

2.2. *Proximate Composition*

To determine the patterns of biologically active compounds accumulation in agro-industrial by-products, it is important to identify their composition and content in separate parts. The proximate composition of the Hess Avocado Seed, in raw, dried, or roasted state is summarized in Table 2. The results indicated that the seed is a good source of dietary protein, with 5.10 g/100 g Fresh Weight (FW) in the raw condition and 5.35 g/100 g FW when the seed is dried or 4.05 g/100 g FW sample when roasted at 160 °C for 60 min. The moisture was reduced in the drying or roasting treatments, leading to different proximate compositions of the samples. The raw avocado seed, besides protein, contains 0.74 g/100 g FW of fat and 49.72 g/100 g FW of carbohydrates and a content of 1.61 g/100 g FW minerals.

Table 2. Proximate composition of raw or conditioned Hass avocado seed expressed as g/100 g (FW).

Sample	Moisture	Protein	Fat	Carbohydrates	Minerals
			g/100 g FW		
Raw	42.83 ± 2.26	5.10 ± 0.003	0.74 ± 0.003	49.72 ± 2.27	1.61 ± 0.01
Dried	29.25 ± 0.16	5.35 ± 0.12	0.65 ± 0.06	62.04 ± 0.18	2.71 ± 0.15
Roasted	0.80 ± 0.01	4.05 ± 0.33	2.12 ± 0.13	90.02 ± 0.41	3.01 ± 0.07

The raw avocado seed will furnish, upon consumption, 225.94 kcal/100 g FW and a higher caloric content was calculated for the dried and roasted samples (275.41/100 g and 395.37 kcal/100 g FW respectively). The currently proposed drink was prepared from 7% of the roasted avocado powder infused with hot water, and thus will have a total caloric amount of 56.48 kcal/100 mL.

The avocado seed is rich in carbohydrates, and thus the further investigation of starch content, which is the most predominant among the carbohydrates, was performed by the Ewers Polarimetric method. In the dried sample, the content is 43.9868 g/100 g FW and in the roasted sample is 48.1192 g/100 g FW. The thermal treatment of the sample might lead to some hydrolysis, increasing the content of starches in comparison with the dried sample. The rest carbohydrates might be represented by dietary fiber [38].

2.3. *Total Polyphenol Content by Folin-Ciocâlteu Method*

As can be seen in Table 3, the highest total polyphenol content was recorded in the crude avocado seed, with a value of 772.90 ± 4.09 mg GAE/100 g FW. The total amount of polyphenols was significantly reduced during processing, while in the drink was as low as 17.55 ± 0.70 mg GAE/100 g FW, given that only 7% of the roasted seed was used for the preparation of the drink. The total polyphenol content of the roasted avocado seed (180 °C/25 min) was of 179.07 ± 4.09 mg GAE/100 g FW; thus, the thermal treatment required in order to obtain organoleptic profiles similar to the coffee alternatives will significantly affect the biologically active compounds which are known to be thermolabile. However, polyphenols are generally more hydrophilic than lipophilic, and thus the proposed hot drink could be a good delivery system for these compounds.

Table 3. The total polyphenol content of raw and conditioned Hass avocado seed and the novel drink.

Samples	Total Phenolic Content (mg GAE/100 g FW)
Crude seed	772.90 ^{a} ± 4.09
Dried seed	279.84 ^{b} ± 2.87
Roasted seed	179.07 ^{c} ± 4.09
The drink	17.55 ^{d} ± 0.70

Identical superscript letters indicate no significant difference ($p > 0.05$).

2.4. *The Total Carotenoid Compounds Content in the Avocado Seed*

The total carotenoid compounds in fruits and vegetables are known to vary because of factors such as genetic variety, maturation stage, or processing conditions. The highest total carotenoid compound in the avocado seed was determined for the dried sample 9228.52 ± 21.20 µg/100 g FW. Similar, but statistically different contents, were determined for the crude and roasted samples (6190.56 ± 14.30 µg/100 g FW and 6534.48 ± 28.30 µg/100 g FW, respectively), as seen in Table 4.

Table 4. The total carotenoid compounds of raw and conditioned Hass avocado seed.

Samples	Total Carotenoid Compounds (µg/100 g FW)
Crude seed	6190.56 [a] ± 14.30
Dried seed	9228.52 [b] ± 21.20
Roasted seed	6534.48 [c] ± 28.30

Identical superscript letters indicate no significant difference ($p > 0.05$).

2.5. Antioxidant Capacity of Avocado Seed and the Drink

The antioxidant capacity reflects the capacity of bioactive compounds to maintain the nutritional and sensorial quality of the product. The highest capacity was observed for the dried seed (95.66 ± 0.566 RSA% DPPH inhibited), but the results registered for the crude and roasted seed were not statistically different (Figure 1). On the other hand, good antioxidant capacity was registered for the proposed drink; even if the roasted avocado seed powder represents 7% of its composition, its antioxidant capacity was 90.27 RSA% DPPH inhibited.

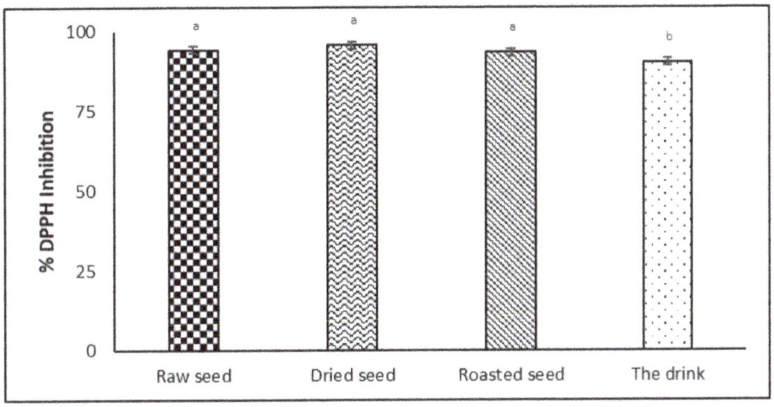

Figure 1. Antioxidant capacity as measured through DPPH assay; identical superscript letters indicate no significant difference ($p > 0.05$).

2.6. The Total Acidity and pH Analysis

The organic acid content present in the crude avocado seed, determined as the titratable acidity, was 0.0538% malic acid equivalents, while in the dried and roasted samples the values were increased to 0.1361% malic acid equivalents and 0.1223% malic acid equivalents respectively. The novel drink registered a total acidity of 0.0268% malic acid equivalents. A pH value of 5.12 ± 0.13 was determined for the roasted avocado seed based drink, this being characteristic for acid drinks.

2.7. The Cytotoxicity Assay of Avocado Seed Drink

Cytotoxicity is considered an important aspect of any new food product or beverage which partially anticipates its health benefits upon consumption. Figures 2 and 3 show the cell viability of the exposed cells to increasing concentrations of the novel roasted avocado seed-based drink.

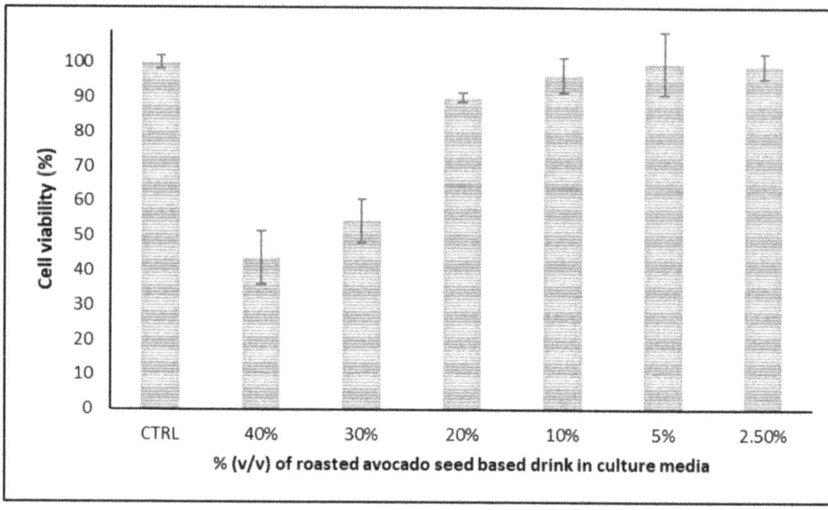

Figure 2. Cell viability of Hs27 cells when treated with the novel roasted avocado seed-based drink.

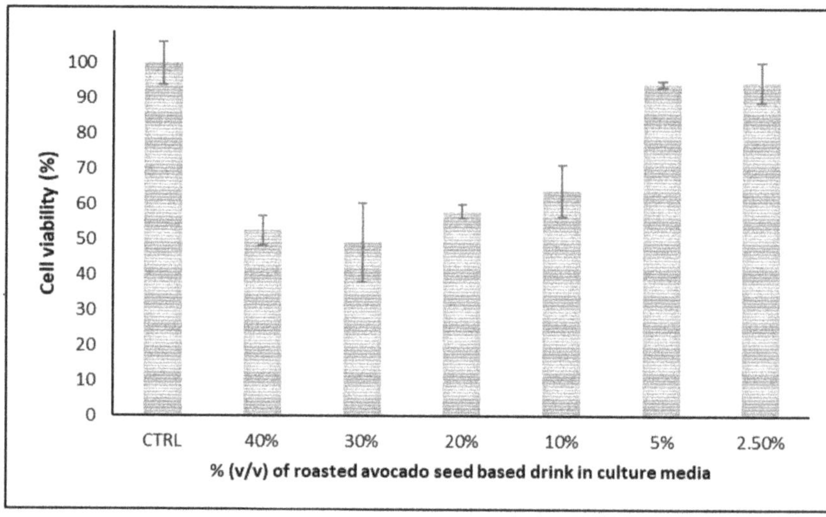

Figure 3. Cell viability of DLD-1 cells when treated with the novel roasted avocado seed-based drink.

In vitro cell viability tests demonstrated less than ≈6% loss in cell viability on both cell lines at the lowest concentration (2.5%) investigated. At the highest concentration (40%), loss in cell viability was observed to be ≈57% for Hs27 cells, whereas the value was ≈47% for the DLD-1 cells at the same concentration. These results showed that the novel drink may decrease the viability of both human fibroblasts (Hs27) and human colorectal adenocarcinoma (DLD-1) cell lines. Furthermore, cell viability experiments demonstrated a dose-dependent response on both cell lines.

3. Discussion

Dried Hass avocado seed powders (65 °C/120–180 min) obtained from ripe or unripe samples were analyzed in terms of proximate composition in another study, in order to be used for preparing extruded snacks [35]. The results of the proximate composition for the ripe seed were slightly different of those presented in the present study, probably due to the longer processing time: the moisture was lower—23.79%, 3.18% protein, 3.33% fibers, 65.62% nitrogen-free extract (carbohydrates), 2.6% ether extract, and 1.51% ashes. This study states the presence of antinutritional factors in the avocado seed (hydrocyanic acid, cyanogenic glycosides, condensed polyphenols and some tannins), which can be eliminated by a cooking treatment [35].

The proximate composition of avocado seed and that of the vitamins were determined in another study after boiling or soaking of the avocado seed. Statistically significant differences ($p < 0.05$) occurred for the content of crude fat, minerals, crude fiber, carbohydrates, and vitamins A, C, and E, during different processing protocols [32]. A content of 0.9% fat and 3.10% protein were determined by other authors in the crude (untreated) Hass Avocado seed [39]. The analysis of individual minerals with the atomic absorption spectrometry method was also performed, revealing high amounts of phosphorus (1000 mg/kg), calcium (533 mg/kg), and magnesium (544 mg/kg); an amount of 1.97% ash was determined, slightly higher than the amount determined in our study [39]. In the Algarvian avocado var. "Hass", lower percentages of protein and a higher amount of fat were determined and the acidity of 2.67 ± 0.17 mg of tartaric acid equivalents/100 g [37]. The starch content of the avocado seed was also investigated by other authors with the Ewers polarimetric method, and the study revealed that higher amounts of starch can be determined in the ripe seed than in the over-ripe seed [39]. Another study reported a starch yield of 42.2% extracted with metabisulfite solution and by producing a dough which was filtered and washed to separate starch from *Daisy* variety [40]. This starches can be further hydrolized using acid or enzyme hydrolysis [41]. The parameters of gelatinization and viscosity of extracted from the avocado seed present restricted dilation, which suggests their possible use in food products which must be heated up at 100 °C [35]. A comparative analysis of the antioxidant capacity of different varieties of avocado seed samples and different extraction condition was carried out in the study of Segovia et al. [42]. In regard to the analysis of total phenolic compounds, the following results were registered for the *Persea americana var.* Hass seed, extracted with Methanol/Water (80:20 *v/v*), 60 °C, namely 9.51 ± 0.16 mg GAE/g [42]. For the ethanolic extracts, the results were of 8.07 ± 0.03 mg GAE/g in the study of Amado et al. [8]. Similar total phenolic compounds were revealed in the Algarvian avocado var. "Hass" (7.04 ± 0.13 mg GAE/g) [43]. Thus, the extraction method applied in the present study is effective in the determination of the total phenolic compounds.

The determination of the total phenolic compounds in a Turkish chicory root, which was roasted for 2 h at 140 °C and grinded to be used as a common coffee alternative, revealed lower contents (between 0.943–13.860 mg GAE/g DW) than those determined in the roasted avocado seed (180 °C/25 min), which were 179.07 ± 4.09 mg GAE/100 g [44]. In the study of Afify et al., the TPC determined for different coffee or teas prepared in hot or cold water, varied between 1.68 ± 0.06 to 2.28 ± 0.06 g GAE/100 g in teas and 1.87 ± 0.07 for a coffee variety, which is lower than the amount determined for the drink prepared from the roasted avocado seed (17.55 ± 0.70 mg GAE/100 g) [45].

Given the high availability, economic advantages, and the abundance of the biologically active compounds, it would be of use to assess the acceptability of the consumers toward food products or beverages having avocado seed as ingredient or spice. It is also a good source of carotenoid compounds, a total amount of 0.97 ± 0.164 mg/100 g fresh weight basis expressed as β-carotene equivalents being determined for the Algarvian avocado var. "Hass". This result is higher than the results exhibited by the crude sample explored in our study. In the current study, slightly higher amounts of total carotenoid compounds were detected in the roasted sample, in comparison with the raw sample, which might be caused by the increase in different isomers of lycopene which are precursors of β -carotene [46].

Another study explored the individual carotenoid compounds determined by HPLC-MS in the Hass Avocado variety originating from Chile, lutein (131.51 μg/g oil extracted from the seed or 2.62 μg/g fresh fresh weight basis) and β-carotene (111.88 μg/g oil extracted from the avocado seed or 2.22 μg/g fresh weight basis) being determined as major carotenoid compounds [47]. Numerous studies explored and demonstrated the in vitro antioxidant and cancer inhibitory activity of avocado seed extracts [2]. Cell viability was assessed using a modified MTT assay on the seed extracts of Hass and Fuerte varieties, and their capacity to inhibit TNFα was assessed in LPS-stimulated RAW 264.7 macrophage culture. None of the tested extracts exhibited cytotoxicity up to 10 μg/mL and the seed exhibited good anti-inflammatory effects after 4 h [48]. In another study, a methanol soluble fraction of the avocado displayed the capability to induce apoptosis and anti-proliferative effects to MCF-7 cell lines [49].

4. Materials and Methods

4.1. Materials

Hass Avocados were purchased from a local market from Cluj Napoca, Romania and were of Columbian origin. All the reagents were of analytical grade.

4.2. Optimization of Roasting of Avocado Seed

The roasting of the avocado seed was performed at different time and temperature intervals to highlight the aroma and to obtain organoleptic characteristics as close as possible to those of coffee or its replacements. The organoleptic analysis was conducted by analyzing the flavor profile of the seed after each time and temperature protocol by a part of the collective of authors who were instructed in regard to the desired organoleptic properties and with previous experience in this [50]. The roasting process was conducted in a Memmert UF55 (Buechenbach, Germany) oven (135 °C for 5–90 min first and for optimization different time temperatures intervals between 160–200 °C, 5–90 min were explored). The temperature in the room was 22–23 °C and the relative humidity was 40–42%.

4.3. The Avocado Seed Powder and the Drink Preparation

The fresh avocado seed contains a high amount of water in the composition and the seeds were naturally dried in a warm airy room, for 5–7 days. The seed is a dicotyledonous and it was kept for 5 min in the oven to facilitate the decortication. The seed was passed through a grater prior to roasting at 180 °C for 25 min. The grater was of stainless steel grade and the side with small holes (diameter 2 mm) was used. After roasting, it was finely ground into a powder using the Retsch RM200 (Haan, Germany) grinding machine set in position 8 (100 rot/min for 20 min).

For the hot drink preparation, water (90 °C ± 5 °C) and a French press were involved until the infusion took place (10 min).

4.4. Proximate Composition Analysis

The chemical compositions including moisture, ash, total carbohydrates, total sugars, crude fat, and protein content were determined for the fresh, dried, and roasted avocado seeds according to AOAC procedures and were expressed in regard to the fresh weight (FW). For moisture analysis, samples were subjected to drying in an oven at 103 ± 2 °C for 3 h, the experiment being repeated until the weight was constant. The samples were cooled in a desiccator for one hour and weighed (AOAC, 1999).

The ash content was determined by calcination at 550 °C of 2 g of probe until a gray ash was obtained, with the removal of the carbon black spots by splashing with water, then the process was continued until a gray or white ash resulted (after 6 h).

The crude protein content of the samples was estimated by the Kjeldahl method.

The crude fat content of the samples was determined by extracting a known weight of powdered samples (3 g) with petroleum ether as a solvent, using the Soxhlet apparatus.

The amount of total carbohydrate was calculated by difference.

The starch content of avocado seed was determined using the Ewers polarimetric method (ISO 10520: 1997) with some modifications [51].

4.5. Total Phenolic Content

The total polyphenols content was assessed using the Folin–Ciocâlteu method [52], slightly modified. An amount of 1 g of sample was mixed with methanol and 0.01% HCl. The obtained extracts were filtered and dried at 35 °C under reduced pressure (Heidolph Rotary Evaporator, Schwabah, Germany). A quantity of 25 µL sample was mixed with 1.8 mL of distilled water and 120 µL Folin-Ciocâlteu reagent in a glass vial. A 7.5% Na_2CO_3 solution prepared in distilled water (340 µL) was added 5 min later to assure basic conditions (pH 10) for the Redox reaction between the phenolic compounds and the Folin-Ciocâlteu reagent. The samples were incubated for 90 min at room temperature. Methanol was used as a control sample. The absorbance at 750 nm was measured using a Shimadzu UV-VIS 1700 spectrophotometer (Shimadzu, Kyoto, Japan). The calibration curve was plotted based on the 0.25, 0.50, 0.75, 1 mg ml^{-1} concentration of gallic acid. The total polyphenol content of the avocado seed was expressed for fresh weight (FW) in Gallic acid equivalents (GAE)—mg GAE·100 g^{-1}.

4.6. Spectroscopic Analysis of the Total Carotenoid Content

To assess the total carotenoid content, carotenoids were extracted from the crude, dried, and roasted avocado seeds, using ethanol: ethyl acetate: petroleum ether (1:1:1, *v/v/v*). Successive extractions were performed. The extracts were combined, filtered, and washed with distilled water, diethyl ether, and a saturated solution of NaOH. The ethereal phase was recovered and subjected to rotary evaporation at 35 °C (Heidolph Rotary Evaporator, Schwabah, Germany). Estimation of carotenoids was spectrophotometrically determined using Shimadzu UV-VIS 1700 set at 450 nm (Shimadzu, Kyoto, Japan).

4.7. The Antioxidant Activity

The antioxidant activity was determined using the 2.2-diphenyl-1-picrylhydrazyl (DPPH) method, according to Mureșan et al. [53]. 10 µL methanolic extract from the avocado seed was mixed with 3.9 mL DPPH methanolic solution (0.025 g/L) and 90 µL distilled water. The mixtures were stirred and maintained properly in the dark for 30 min. The absorbance of the samples was measured at 515 nm (Shimadzu 1700 UVVIS, Kyoto, Japan) against a methanol blank. The positive control was prepared using a gallic acid solution (0.5 mg/mL). The negative control was prepared using methanol. Results were expressed as percent over standard DPPH absorbance according to the following equation:

$$\text{RSA}\ [\%] = \frac{A_{DPPH} - A_P}{A_{DPPH}} \bullet 100$$

where: RSA [%]—Radical Scavenging Activity; A_{DPPH}—the absorbance of DPPH solution with methanol; A_P—the absorbance of DPPH solution after 30 min incubation with sample.

4.8. The Total Acidity and pH Analysis

The total acidity was performed by neutralization with sodium hydroxide solution (0.1 N) in the presence of fenolftalein as indicator. The results were expressed in g malic acid equivalents/100 g. Titratable acidity calculation was carried out using the formula:

$$\text{Acidity}\ \%\ (\text{malic acid}) = \frac{V \bullet 0.0067}{m} \bullet 100,$$

where 'V' is the volume of NaOH solution 0.1N, 'm' is the weight of the sample, and '0.0067' g of malic acid corresponds to 1 mL NaOH 0.1 N.

pH analysis was also carried out on the novel drink. The determination was based on the property of indicators contained by the pH paper to change their color in the presence of hydrogen ions. To determine the pH of the novel drink, the pH meter Mettler Toledo (Columbus, OH, USA) was employed.

4.9. Cytotoxicity Assay

The cytotoxicity assay of the avocado seed drink was performed using human fibroblasts Hs27 (ATCC® CRL-1634™, Manassas, VA, USA) and human colorectal adenocarcinoma DLD-1 (ATCC® CCL-221™, Manassas, VA, USA) cell lines. The cells were cultured according to standard conditions. The potential cytotoxicity of avocado seed drink was assessed with (4,5-dimethylthiazol-2-yl)-2,5-diphenyltetrazolium bromide (MTT) assay. In order to obtain cell suspensions, the cells were treated with 0.25% trypsin-EDTA, and after centrifugation (1500 rpm for 5 min), 1×10^4 cells/well were seeded on 96-well plates in 200 µL complete culture medium. After 24 h, 80 µL (40% *v/v*), 60 µL (30% *v/v*), 40 µL (20% *v/v*), 10 µL (5% *v/v*), and 5 µL (2.5% *v/v*) of avocado seed drink were added, while removing the same volume of culture media, resulting in a total final volume of 200 µL/well. Control samples were represented by untreated cells. Each experimental condition was performed in triplicate. Cell proliferation analysis was performed after 24 h. After 24 h, the medium was removed and 100 µL of 1 mg/mL MTT solution (Sigma-Aldrich, St. Louis, MO, USA) was added. After 4 h of incubation at 37 °C in dark, the MTT solution was removed from each well and 150 µL of DMSO (dimethyl sulfoxide) solution (Fluka, Buchs, Switzerland) was added. Spectrophotometric readings at 450 nm were performed with a BioTek Synergy 2 microplate reader (Winooski, VT, USA). Data are shown as percentage of cell viability.

4.10. Statistics

Statistical differences were obtained through an analysis of variance (ANOVA) followed by Tukey's multiple comparison test at 95% confidence level ($p \leq 0.05$).

5. Conclusions

The avocado seed has been explored lately for various applications due to its composition rich in bioactive compounds. Numerous processing techniques such as boiling, dryingor extrusion, have been explored so far as conditioning methods, due to the presence of anti-nutritive factors in its composition. The current study evaluated different time-temperature protocols for the roasting of the avocado seed, along with drying, and the flavor and color modification were assessed. We proposed the valorization of the roasted (180 °C/25 min) avocado seed powder in a hot drink, obtained by creating an infusion with 7% of the powder and hot water.

The raw or conditioned (dried or roasted) Hass avocado seeds were examined in terms of proximate composition and bioactive compounds. The seed possess a high amount of carbohydrates, including dietary fibers and between 4–5% protein, depending on the conditioning process applied (drying or roasting). The total polyphenolic content of the avocado seed was reduced during the conditioning, while the acidity and total carotenoid compound were significantly increased. The novel drink exhibited a high antioxidant capacity of 90.27 RSA% DPPH inhibited, which might be due to the presence of carotenoid compounds or flavonoids.

However, the novel drink exhibited a lower concentration of the total polyphenolic compounds in comparison with the raw or conditioned seed (only 17.55 ± 0.70 mg GAE/100 g in the drink compared with 179.07 ± 4.09 mg GAE/100 g in the roasted avocado seed), mostly because the drink has a roasted avocado seed powder concentration as low as 7%. This is a higher content that what was previously registered for coffee or coffee surrogates in other studies. The cytotoxic properties of the novel drink based on roasted avocado seed were also demonstrated, because when it was applied in a concentration of 40% on DLD-1 cells and Hs27 cells during the MTT cytotoxicity assay, it affected the viability of the cells.

Author Contributions: Conceptualization, A.E.M.; data curation, A.P.; formal analysis, A.P., A.E.M., A.E.T., R.A.M., E.P. and C.C.; investigation, V.M. and C.C.; methodology, A.P. and A.E.M.; project administration, A.E.M.; supervision, A.E.M., V.M. and C.C.; writing—original draft, A.P.; writing—review and editing, V.M. All authors have read and agreed to the published version of the manuscript.

Funding: This work was supported by a grant of the Ministry of Research, Innovation and Digitization, CNCS/CCCDI—UEFISCDI, project number PN-III-P1-1.1-PD-2019-1108, within PNCDI III and the publication was supported by funds from the National Research Development Projects to finance excellence (PFE)-14/2022-2024 granted by the Romanian Ministry of Research and Innovation.

Institutional Review Board Statement: Not applicable.

Informed Consent Statement: Not applicable.

Data Availability Statement: Not applicable.

Conflicts of Interest: The authors declare no conflict of interest.

References

1. Bahru, T.B.; Tadele, Z.H.; Ajebe, E.G. A review on avocado seed: Functionality, composition, antioxidant and antimicrobial properties. *Chem. Sci. Int. J.* **2019**, *27*, 45609. [CrossRef]
2. Dabas, D.; Elias, R.J.; Ziegler, G.R.; Lambert, J.D. In vitro antioxidant and cancer inhibitory activity of a colored avocado seed extract. *Int. J. Food Sci.* **2019**, *2019*, 6509421. [CrossRef] [PubMed]
3. Segovia, F.J.; Corral-Pérez, J.J.; Almajano, M.P. Avocado seed: Modeling extraction of bioactive compounds. *Ind. Crops Prod.* **2016**, *85*, 213–220. [CrossRef]
4. Yepes-Betancur, D.P.; Márquez-Cardozo, C.J.; Cadena-Chamorro, E.M.; Martinez-Saldarriaga, J.; Torres-León, C.; Ascacio-Valdes, A.; Aguilar, C.N. Solid-state fermentation—Assisted extraction of bioactive compounds from hass avocado seeds. *Food Bioprod. Process.* **2021**, *126*, 155–163. [CrossRef]
5. Brzezińska, R.; Górska, A.; Gotowicka, K.; Bryś, J.; Ostrowska-Ligęza, E.; Wirkowska-Wojdyła, M. Quality assessment of avocado pulp oils during storage. *Multidiscip. Digit. Publ. Inst. Proc.* **2020**, *70*, 14.
6. Pieterse, Z.; Jerling, J.; Stonehouse, W. Avocados (monounsaturated fatty acids), weight loss and serum lipids. *S. Afr. Avocado Grow. Assoc. Yearb.* **2003**, *26*, 65–71.
7. Dreher, M.L.; Cheng, F.W.; Ford, N.A. A Comprehensive Review of Hass Avocado Clinical Trials, Observational Studies, and Biological Mechanisms. *Nutrients* **2021**, *13*, 4376. [CrossRef]
8. Amado, D.A.V.; Helmann, G.A.B.; Detoni, A.M.; de Carvalho, S.L.C.; De Aguiar, C.M.; Martin, C.A.; Tiuman, T.S.; Cottica, S.M. Antioxidant and antibacterial activity and preliminary toxicity analysis of four varieties of avocado (*Persea americana* Mill.). *Braz. J. Food Technol.* **2019**, *22*, e2018044. [CrossRef]
9. Chaves, M.; Mendonça, C.R.B.; Borges, C.; Porcu, O. Preparation of whole cookie using avocado pulp flour and oil. *Bol. Cent. Pesqui. Process. Aliment.* **2013**, *31*, 215–226.
10. Duarte, P.F.; Chaves, M.A.; Borges, C.D.; Mendonça, C.R.B. Avocado: Characteristics, health benefits and uses. *Ciênc. Rural.* **2016**, *46*, 747–754. [CrossRef]
11. Heskey, C.; Oda, K.; Sabaté, J.J.N. Avocado Intake, and Longitudinal Weight and Body Mass Index Changes in an Adult Cohort. *Nutrients* **2019**, *11*, 691. [CrossRef]
12. Da Silva, G.G.; Pimenta, L.P.S.; Melo, J.O.F.; Mendonça, H.d.O.P.; Augusti, R.; Takahashi, J.A. Phytochemicals of Avocado Residues as Potential Acetylcholinesterase Inhibitors, Antioxidants, and Neuroprotective Agents. *Molecules* **2022**, *27*, 1892. [CrossRef]
13. Tesfaye, T.; Ayele, M.; Gibril, M.; Ferede, E.; Limeneh, D.Y.; Kong, F. Beneficiation of avocado processing industry by-product: A review on future prospect. *Curr. Res. Green Sustain. Chem.* **2022**, *5*, 100253. [CrossRef]
14. Rodríguez-Carpena, J.-G.; Morcuende, D.; Andrade, M.-J.; Kylli, P.; Estévez, M. Avocado (*Persea americana* Mill.) phenolics, in vitro antioxidant and antimicrobial activities, and inhibition of lipid and protein oxidation in porcine patties. *J. Agric. Food Chem.* **2011**, *59*, 5625–5635. [CrossRef]
15. Dabas, D.; Shegog, M.R.; Ziegler, R.G.; Lambert, D.J. Avocado (*Persea americana*) seed as a source of bioactive phytochemicals. *Curr. Pharm. Des.* **2013**, *19*, 6133–6140. [CrossRef]
16. Soledad, C.-P.T.; Paola, H.-C.; Carlos Enrique, O.-V.; Israel, R.-L.I.; GuadalupeVirginia, N.-M.; Raúl, Á.-S. Avocado seeds (*Persea americana* cv. *Criollo* sp.): Lipophilic compounds profile and biological activities. *Saudi J. Biol. Sci.* **2021**, *28*, 3384–3390. [CrossRef]
17. Do, D.T.B.; Bui, T.H.; Phan, D.T.A. *Persea Americana* Mill seed extracts: Understanding insights into the antioxidant and antityrosinase activities and effects on preserving qualities of whiteleg shrimp (*Litopenaus vannamei*) during refrigerated storage. *Food Chem.* **2022**, *373*, 131469. [CrossRef]
18. Melgar, B.; Dias, M.I.; Ciric, A.; Sokovic, M.; Garcia-Castello, E.M.; Rodriguez-Lopez, A.D.; Barros, L.; Ferreira, I.C.R.F. Bioactive characterization of *Persea americana* Mill. by-products: A rich source of inherent antioxidants. *Ind. Crops Prod.* **2018**, *111*, 212–218. [CrossRef]
19. Leite, J.J.G.; Brito, É.H.S.; Cordeiro, R.A.; Brilhante, R.S.N.; Sidrim, J.J.C.; Bertini, L.M.; Morais, S.M.d.; Rocha, M.F.G. Chemical composition, toxicity and larvicidal and antifungal activities of *Persea americana* (avocado) seed extracts. *Rev. Soc. Bras. Med. Trop.* **2009**, *42*, 110–113. [CrossRef]
20. Figueroa, J.G.; Borrás-Linares, I.; Lozano-Sánchez, J.; Segura-Carretero, A.J. Comprehensive characterization of phenolic and other polar compounds in the seed and seed coat of avocado by HPLC-DAD-ESI-QTOF-MS. *Food Res. Int.* **2018**, *105*, 752–763. [CrossRef]

21. Kristanti, C.D.; Simanjuntak, F.P.J.; Dewi, N.K.P.A.; Tianri, S.V.; Hendra, P.J. Community. Anti-inflammatory and Analgesic Activities of Avocado Seed (*Persea americana* Mill.). *J. Pharm. Sci. Community* **2017**, *14*, 104–111. [CrossRef]
22. Jiménez-Arellanes, A.; Luna-Herrera, J.; Ruiz-Nicolás, R.; Cornejo-Garrido, J.; Tapia, A.; Yépez-Mulia, L.J.B.c. Antiprotozoal and antimycobacterial activities of *Persea americana* seeds. *BMC Complement. Altern. Med.* **2013**, *13*, 109. [CrossRef] [PubMed]
23. Ramos-Jerz, M.d.R.; Villanueva, S.; Jerz, G.; Winterhalter, P.; Deters, A.M. *Persea americana* Mill. seed: Fractionation, characterization, and effects on human keratinocytes and fibroblasts. *Evid. -Based Complement. Altern. Med.* **2013**, *2013*, 391247. [CrossRef] [PubMed]
24. Alkhalaf, M.I.; Alansari, W.S.; Ibrahim, E.A.; ELhalwagy, M.E. Anti-oxidant, anti-inflammatory and anti-cancer activities of avocado (*Persea americana*) fruit and seed extract. *J. King Saud Univ. Sci.* **2019**, *31*, 1358–1362. [CrossRef]
25. Ezejiofor, A.N.; Okorie, A.; Orisakwe, O.E. Hypoglycaemic and tissue-protective effects of the aqueous extract of *Persea americana* seeds on alloxan-induced albino rats. *Malays. J. Med. Sci.* **2013**, *20*, 31.
26. Athaydes, B.R.; Alves, G.M.; De Assis, A.L.E.M.; Gomes, J.V.D.; Rodrigues, R.P.; Campagnaro, B.P.; Nogueira, B.V.; Silveira, D.; Kuster, R.M.; Pereira, R.R.; et al. Avocado seeds (*Persea americana* Mill.) prevents indomethacin-induced gastric ulcer in mice. *Food Res. Int.* **2019**, *119*, 751–760. [CrossRef]
27. Bhuyan, D.J.; Alsherbiny, M.A.; Perera, S.; Low, M.; Basu, A.; Devi, O.A.; Barooah, M.S.; Li, C.G.; Papoutsis, K. The Odyssey of Bioactive Compounds in Avocado (*Persea americana*) and Their Health Benefits. *Antioxidants* **2019**, *8*, 426. [CrossRef]
28. Lu, Q.-Y.; Arteaga, J.R.; Zhang, Q.; Huerta, S.; Go, V.L.W.; Heber, D. Inhibition of prostate cancer cell growth by an avocado extract: Role of lipid-soluble bioactive substances. *J. Nutr. Biochem.* **2005**, *16*, 23–30. [CrossRef]
29. Rodríguez-Carpena, J.G.; Morcuende, D.; Estévez, M. Avocado by-products as inhibitors of color deterioration and lipid and protein oxidation in raw porcine patties subjected to chilled storage. *Meat Sci.* **2011**, *89*, 166–173. [CrossRef]
30. Dabas, D.; Elias, R.J.; Lambert, J.D.; Ziegler, G.R. A colored avocado seed extract as a potential natural colorant. *J. Food Sci.* **2011**, *76*, C1335–C1341. [CrossRef]
31. Padilla-Camberos, E.; Martínez-Velázquez, M.; Flores-Fernández, J.M.; Villanueva-Rodríguez, S. Acute toxicity and genotoxic activity of avocado seed extract (*Persea americana* Mill., cv Hass). *Sci. World J.* **2013**, *2013*, 245828. [CrossRef]
32. Talabi, J.Y.; Osukoya, O.A.; Ajayi, O.; Adegoke, G.O. Nutritional and antinutritional compositions of processed Avocado (*Persea americana* Mill) seeds. *Asian J. Plant Sci. Res.* **2016**, *6*, 6–12.
33. Alissa, K.; Hung, Y.-C.; Hou, C.Y.; Lim, G.C.W.; Ciou, J.-Y. Developing new health material: The utilization of spray drying technology on avocado (*Persea americana* Mill.) seed powder. *Foods* **2020**, *9*, 139. [CrossRef]
34. Puspitasari, M.; Hasri, R.N.; Wahyuni, T. The Substitution of Avocado Seed Flour to Rice Flour in the Manufacture of Traditional Palembang Food Gandus Cake. In Proceedings of the 1st International Conference on Health, Social Sciences and Technology (ICOHSST 2020), Palembang, Indonesia, 19 April 2021; pp. 37–40.
35. Olaeta, J.; Schwartz, M.; Undurraga, P.; Contreras, S. Use of Hass avocado (*Persea americana* Mill) seed as a processed product. In Proceedings of the 6th World Avocado Congress, Viña del Mar, Chile, 12–16 November 2007; pp. 1–8.
36. Ifesan, B.; Olorunsola, B.; Ifesan, B.T. Nutritional composition and acceptability of candy from avocado seed (*Persea americana*). *Int. J. Agric. Innov. Res.* **2015**, *3*, 1732–1735.
37. Mahawan, M.A.; Tenorio, M.F.N.; Gomez, J.A.; Bronce, R.A. Characterization of flour from avocado seed kernel. *Asia Pac. J. Multidiscip. Res.* **2015**, *3*, 34–40.
38. Barbosa-Martín, E.; Chel-Guerrero, L.; González-Mondragón, E.; Betancur-Ancona, D. Chemical and technological properties of avocado (*Persea americana* Mill.) seed fibrous residues. *Food Bioprod. Process.* **2016**, *100*, 457–463. [CrossRef]
39. Diana, Y.; Carlos, M.; Edith, C. Technology. Effect of maturity state of avocado (*Persea americana* Mill. cv. Hass) on seed characteristics. *Adv. J. Food Sci. Technol.* **2018**, *16*, 301–306. [CrossRef]
40. Dos Santos, D.M.; Ascheri, D.P.R.; De Lacerda Bukzem, A.; Morais, C.C.; Carvalho, C.W.P.; Ascheri, J.L.R. Physicochemical properties of starch from avocado seed (*Persea Americana* Mill). *Bol. Cent. Pesqui. Processl. Aliment.* **2017**, *34*. [CrossRef]
41. Maryam, M.; Kasim, A.; Santosa. Utilization Starch of Avocado Seed (*Persea Americana* Mill.) as a Raw Material for Dextrin. *J. Food Sci. Eng.* **2016**, *6*, 32–37. [CrossRef]
42. Segovia, F.J.; Hidalgo, G.I.; Villasante, J.; Ramis, X.; Almajano, M.P. Avocado seed: A comparative study of antioxidant content and capacity in protecting oil models from oxidation. *Molecules* **2018**, *23*, 2421. [CrossRef]
43. Ferreira da Vinha, A.; Silva, C.S.; Soares, M.D.O.; Barreira, S. Avocado and its by-products: Natural sources of nutrients, phytochemical compounds and functional properties. *Curr. Res. Agric. Food Sci.* **2020**, *1*, 82–96.
44. Başaran, U.; Gülümser, E.; Yaman, C.; Doğrusöz, M.Ç.; Mut, H. Antioxidants and Mineral Contents of Chicory as Coffee Additive. *Turk. J. Agric. Food Sci. Technol.* **2021**, *9*, 217–223. [CrossRef]
45. Afify, A.E.-M.M.; Shalaby, E.A.; El-Beltagi, H. Antioxidant Activity of Aqueous Extracts of Different Caffeine Products. *Not. Bot. Horti Agrobot.* **2011**, *39*, 117–123. [CrossRef]
46. D'Evoli, L.; Lombardi-Boccia, G.; Lucarini, M. Influence of heat treatments on carotenoid content of cherry tomatoes. *Foods* **2013**, *2*, 352–363. [CrossRef]
47. Flores, M.; Ortiz-Viedma, J.; Curaqueo, A.; Rodriguez, A.; Dovale-Rosabal, G.; Magaña, F.; Vega, C.; Toro, M.; López, L.; Ferreyra, R.; et al. Preliminary Studies of Chemical and Physical Properties of Two Varieties of Avocado Seeds Grown in Chile. *J. Food Qual.* **2019**, *2019*, 3563750. [CrossRef]

48. Tremocoldi, M.A.; Rosalen, P.L.; Franchin, M.; Massarioli, A.P.; Denny, C.; Daiuto, É.R.; Paschoal, J.A.R.; Melo, P.S.; Alencar, S.M. Exploration of avocado by-products as natural sources of bioactive compounds. *PLoS ONE* **2018**, *13*, e0192577. [CrossRef]
49. Widiyastuti, Y.; Pratiwi, R.; Riyanto, S.; Wahyuono, S. Cytotoxic activity and apoptosis induction of avocado *Persea americana* Mill. seed extract on MCF-7 cancer cell line. *Indones. J. Biotechnol.* **2018**, *23*, 61–67. [CrossRef]
50. Marc (Vlaic), R.A.; Niculae, M.; Páll, E.; Mureșan, V.; Mureșan, A.; Tanislav, A.; Pușcaș, A.; Mureșan, C.C.; Cerbu, C. Red Oak (*Quercus rubra* L.) Fruits as Potential Alternative for Cocoa Powder: Optimization of Roasting Conditions, Antioxidant, and Biological Properties. *Forests* **2021**, *12*, 1088. [CrossRef]
51. Cerbu, E.; Muste, S.; Borşa, A.; Man, S.; Mureşan, V. Relationship between the polarimetric determination of starch and the starch iodine index in apple variety Jonathan. *Bull. UASVM Agric.* **2011**, *68*, 533.
52. Singleton, V.L.; Orthofer, R.; Lamuela-Raventós, R.M. [14] Analysis of total phenols and other oxidation substrates and antioxidants by means of folin-ciocalteu reagent. *Methods Enzymol.* **1999**, *299*, 152–178.
53. Muresan, A.E.; Muste, S.; Borsa, A.; Vlaic, R.A.; Muresan, V. Evaluation of physical-chemical indexes, sugars, pigments and phenolic compounds of fruits from three apple varieties at the end of storage period. *Bull. UASVM Food Sci. Technol.* **2014**, *71*, 45–50. [CrossRef]

plants

MDPI

Article

Antidiarrheal and Cardio-Depressant Effects of *Himalaiella heteromalla* (D.Don) Raab-Straube: In Vitro, In Vivo, and In Silico Studies

Fatima Saqib [1,*], Faisal Usman [1], Shehneela Malik [1], Naheed Bano [2], Najm Ur-Rahman [3], Muhammad Riaz [3], Romina Alina Marc (Vlaic) [4,*] and Crina Carmen Mureşan [4]

1 Faculty of Pharmacy, Bahauddin Zakariya University, Multan 60000, Pakistan; faisal.usman@bzu.edu.pk (F.U.); shehneelamalik1@gmail.com (S.M.)
2 Faculty of Veterinary and Animal Sciences, MNS-University of Agriculture, Multan 60000, Pakistan; bnaheed61@gmail.com
3 Department of Pharmacy, Shaheed Benazir Bhutto University, Sheringal 18050, Pakistan; najm@sbbu.edu.pk (N.U.-R.); pharmariaz@gmail.com (M.R.)
4 Food Engineering Department, Faculty of Food Science and Technology, University of Agricultural Sciences and Veterinary Medicine, 400372 Cluj-Napoca, Romania; crina.muresan@usamvcluj.ro
* Correspondence: fatima.saqib@bzu.edu.pk (F.S.); romina.vlaic@usamvcluj.ro (R.A.M.)

Abstract: *Himalaiella heteromalla* (D.Don) Raab-Straube is a commonly used remedy against various diseases. Crude extract and fractions of *H. heteromalla* were investigated for a gastrointestinal, bronchodilator, cardiovascular, and anti-inflammatory activities. *H. heteromalla* crude extract (Hh.Cr) relaxed spontaneous contractions and K^+ (80 mM)-induced contraction in jejunum tissue dose-dependently. The relaxation of K^+ (80 mM) indicates the presence of Ca^{++} channel blocking (CCB) effect, which was further confirmed by constructing calcium response curves (CRCs) as they caused rightward parallel shift of CRCs in a manner comparable to verapamil, so the spasmolytic effect of Hh.Cr was due to its CCB activity. Application of Hh.Cr on CCh (1 μM) and K^+ (80 mM)-induced contraction in tracheal preparation resulted in complete relaxation, showing its bronchodilator effect mediated through Ca^{++} channels and cholinergic antagonist activity. Application of Hh.Cr on aortic preparations exhibited vasorelaxant activity through angiotensin and α-adrenergic receptors blockage. It also showed the cardio suppressant effect with negative chronotropic and inotropic response in paired atrium preparation. Similar effects were observed in in vivo models, i.e., decreased propulsive movement, wet feces, and inhibition of edema formation.

Keywords: antidiarrheal; calcium ion channel; cardio-depressant; *Himalaiella heteromalla*

Citation: Saqib, F.; Usman, F.; Malik, S.; Bano, N.; Ur-Rahman, N.; Riaz, M.; Marc (Vlaic), R.A.; Mureşan, C.C. Antidiarrheal and Cardio-Depressant Effects of *Himalaiella heteromalla* (D.Don) Raab-Straube: In Vitro, In Vivo, and In Silico Studies. *Plants* **2022**, *11*, 78. https://doi.org/10.3390/plants11010078

Academic Editor: Antonella Smeriglio

Received: 7 December 2021
Accepted: 20 December 2021
Published: 27 December 2021

1. Introduction

Himalaiella heteromalla (D.Don) Raab-Straube (*Asteraceae*), commonly known as Batula, is found in low-temperature regions of Asia, Europe, and North America [1]. *Himalaiella heteromalla* is a rich source of chlorojanerin, arctigenin [2] glycosides, alkaloids, terpenoids, saponins, flavonoids, sesquiterpene lactones, and arctiin [3]. Gao et al. [4] and Kang et al. [5] reported arctigenin and its glycoside, arctiin, have anti-inflammatory activities by inhibiting iNOS and exerting vasodilation effect, while Hayashi et al. [6] reported the anti-viral activity against influenza A virus.

Traditionally, *H. heteromalla* is used in herbal products to treat fever, menstruation, circulation, pain, and rheumatic arthritis [7]. It is used in wounds, cuts, and fever [8]. The leaf paste and mustard oil mixture are used for wounds and leukoderma. It has carminative property, used for coeliac diseases [9,10]. It is used to remedy burning parts of the body, menstrual problems, piles, psoriasis, rheumatoid arthritis, cardiotonic cough with cold, and altitude sickness, and provide anticancer and anti-fatigue actions [11]. It is used as an anti-inflammatory and prevents ischemic stroke [3]. Therefore, *H. heteromalla* was investigated in in vitro, in vivo, and in silico models as a possible tool to treat gastrointestinal, cardiovascular, respiratory, and inflammatory ailments.

2. Results

2.1. Phytochemical Analysis of *Himalaiella heteromalla*

The preliminary phytochemical analysis of Hh.Cr confirmed the presence of glycosides, saponins, alkaloids, and flavonoids.

2.2. HPLC Separation of Phenolic Acids and Flavonoids

The separation factor and resolution of all separated compounds were >1.0 and >1.5, respectively. The reproducibility of separate components was also good with RSD < 2% (run to run) and 2.7% (day to day) [12]. The HPLC chromatograms for the identified phenolic and flavonoid compounds are shown in Figure 1, and Table 1 presents the phenolic and flavonoid compounds identified in Hh.Cr. The most abundant phenolic compounds were gallic acid (184.98 μg/g), hydroxybenzoic acid (6.8 μg/g), and vanillic acid (8.1 μg/g); while the identified flavonoid compound was catechin (160.37 μg/g).

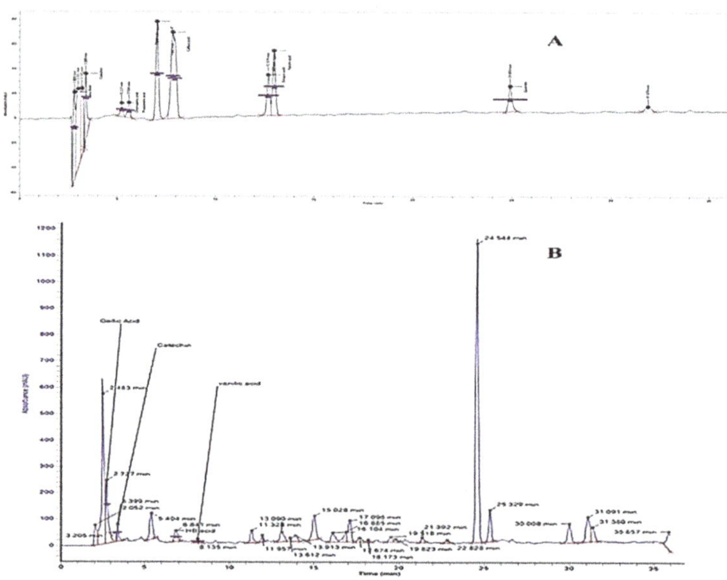

Figure 1. HPLC chromatogram of (**A**) standard phenolic compounds (**B**) Hh.Cr.

Table 1. Phenolic and flavonoid compounds of Hh.Cr.

Sr.No.	Compound	Retention Time (min)	Concentration (μg/g)
1.	Gallic Acid	2.7	184.98
2.	Catechin	3.3	160.37
3.	HB acid	6.8	22.80
4.	Vanilic acid	8.1	9.08

2.3. Effect on Jejunum Preparations

The Hh.Cr and its ethyl acetate (Hh.Ea) fraction showed the relaxant effect on exposure to the rhythmic contraction of jejunum preparations in organ bath within concentration range 0.01 to 0.3 mg/mL with EC_{50} 0.06 mg/mL (95% CI: 0.045–0.080 mg/mL; n = 5) and 0.01 to 0.1 mg/mL with EC_{50} 0.032 mg/mL (95% CI: 0.021–0.51 mg/mL; n = 5), respectively similar to verapamil with EC_{50} 0.42 μM (95% CI: 0.22–1.27), whereas aqueous fraction *H. heteromalla* failed to complete relaxation of spontaneous contractions of jejunum preparations. Hh.Cr and Hh.Ea also caused a complete relaxation of K^+ (80 mM) induced spastic contractions at 1 mg/mL with EC_{50} 0.13 mg/mL (95% CI: 0.088–0.220 mg/mL; n = 5) and 0.3 mg/mL with EC_{50} 0.06 mg/mL (95% CI: 0.045–0.0.089 mg/mL; n = 5), respectively similar to verapamil with EC_{50} 0.251 μM (95% CI: 0.082–0.784). Furthermore, Hh.Cr showed a rightward shift of calcium CRCs which confirm the presence calcium ion channel blockade activity in Hh.Cr, similar to verapamil (Figures 2 and 3).

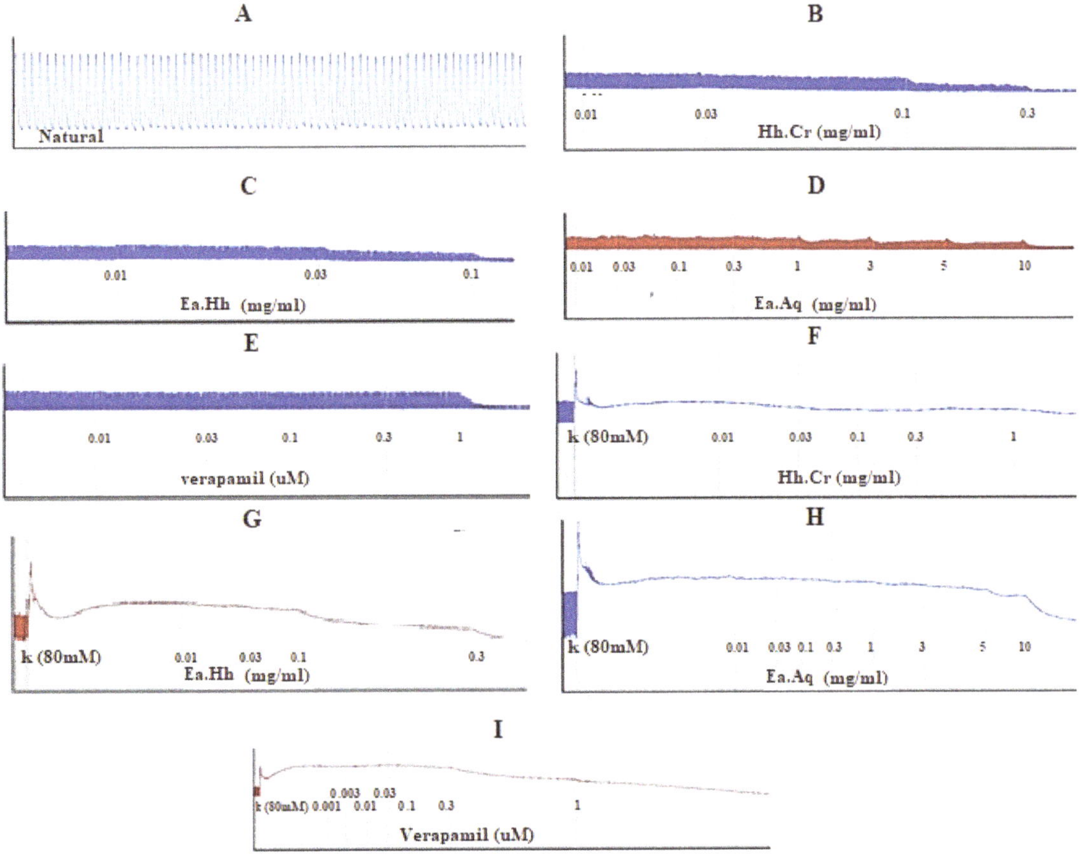

Figure 2. (**A**) Control spontaneous contraction. Effect of (**B**) crude extract (Hh.Cr), (**C**) Ethyl acetate fraction (Ea.Hh), (**D**) Aqueous fraction (Ea.Aq), and (**E**) verapamil on spontaneous. Effect of (**F**) crude extract (Hh.Cr), (**G**) Ethyl acetate fraction (Ea.Hh), (**H**) Aqueous fraction (Ea.Aq), and (**I**) verapamil on K^+ Induced Contraction on rabbit jejunum preparations.

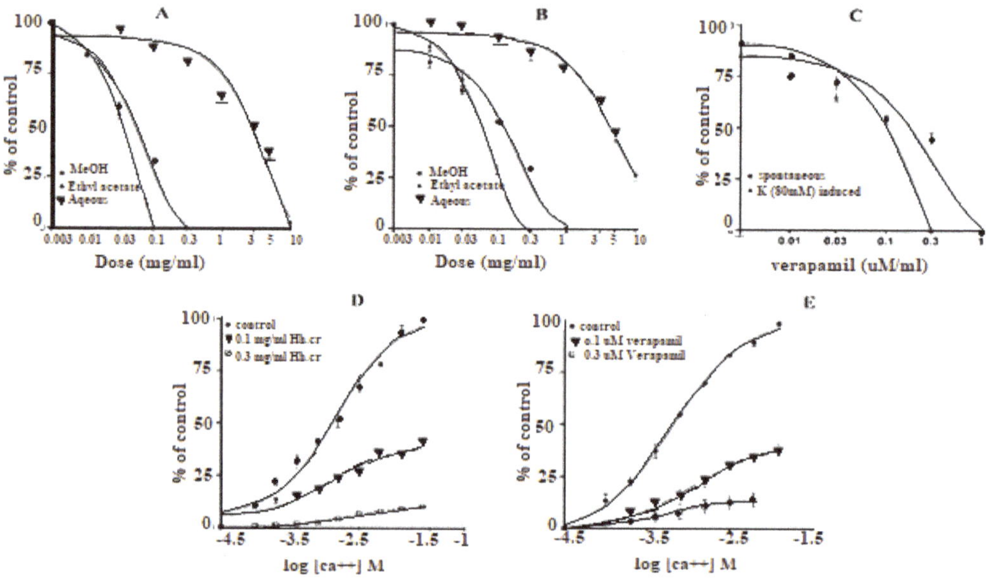

Figure 3. Effect of crude extract (Hh.Cr) Ethyl acetate fraction (Ea.Hh) and aqueous fraction (Ea.Aq) on (**A**) spontaneous contraction and (**B**) K^+ (80 mM)-induce Contraction on rabbit jejunum preparations. (**C**) Effect of verapamil on spontaneous and K^+ induced contraction on rabbit jejunum preparations. Dose–response curves of Ca^{++} in the presence and absence of (**D**) Hh.Cr (**E**) verapamil in the isolated rabbit jejunum preparations. Values are expressed as mean ± SEM.

2.4. Effect on Tracheal Preparations

The Hh.Cr and its ethyl acetate (Hh.Ea) fraction showed the relaxant effect on tracheal preparations, when exposed K^+ (80 mM) and CCh (1 µM)-induced contractions. Hh.Cr and its Hh.Ea fraction relaxed the K^+ (80 mM) induce contractions at 0.3 mg/mL with EC_{50} 0.19 mg/mL (95% CI: 0.099–0.452; n = 5) and 0.1 mg/mL with EC_{50} 0.042 mg/mL (95% CI: 0.024–0.072 mg/mL; n = 5), respectively. Hh.Cr and its Hh.Ea fraction also relaxed the CCh (1 µM) induce contractions at 1 mg/mL with EC_{50} 0.23 mg/mL (95% CI: 0.158–0.357 mg/mL; n = 5) and 0.3 mg/mL with EC_{50} 0.155 mg/mL (95% CI: 0.076–0.302 mg/mL; n = 5), respectively. Similarly, verapamil also caused relaxation of K^+ (80 mM) and CCh (1 µM) induced contractions with respective EC_{50} 0.82 µM (95% CI: 0.82–0.82 µM) and EC_{50} values of 2.35 µM (95% CI: 02.32–2.39 µM). The aqueous fraction (Hh.Aq) exerted partially relaxation of K^+ (80 mM) and CCh (1 µM)-induced contractions on tracheal preparation (Figure 4).

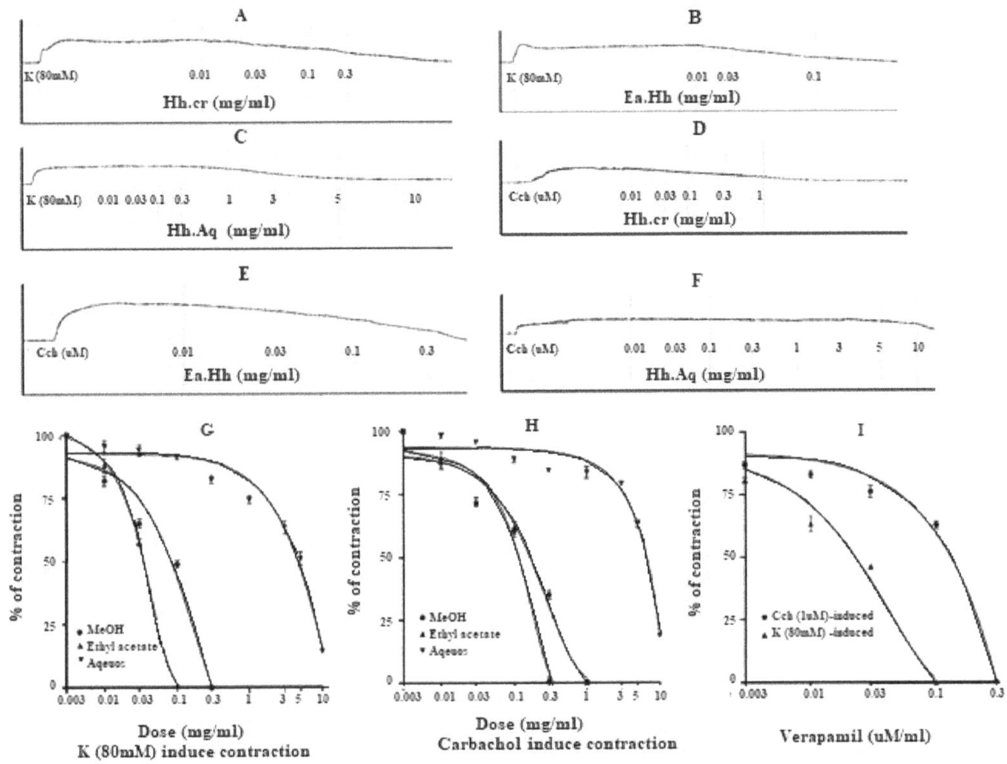

Figure 4. Effect of (**A**) crude extract (Hh.Cr) (**B**) Ethyl acetate fraction (Ea.Hh) (**C**) Aqueous fraction (Ea.Aq) on K^+ Induce Contraction and Effect of (**D**) crude extract (Hh.Cr) (**E**) Ethyl acetate fraction (Ea.Hh) (**F**) Aqueous fraction (Ea.Aq) on CCh-induced contraction on rabbit tracheal preparations. Effect of crude extract (Hh.Cr) ethyl acetate fraction (Ea.Hh) and aqueous fraction (Ea.Aq) of on (**G**) K^+ (80 mM) Induce Contraction and (**H**) CCh-induced contraction on tracheal preparations. (**I**) Effect of verapamil on CCh1 µM and K^+ induced contraction on rabbit tracheal preparations. Values are expressed as mean ± SEM.

2.5. Effect on Aortic Preparations

The crude extract (Hh.Cr) and its ethyl acetate (Hh.Ea) fraction showed the relaxant effect on aortic preparations, when exposed K^+ (80 mM) and PE (1 µM) induced contractions. Hh.Cr and its Hh.Ea fraction relaxed the K^+ (80 mM) induce contractions at 3 mg/mL with EC_{50} 2.88 mg/mL (95% CI: 2.106–4.156 mg/mL; n = 5) and 1 mg/mL with EC_{50} 0.148 mg/mL (95% CI: 0.09491–0.2332 mg/mL; n = 5), respectively. Hh.Cr and its Hh.Ea fraction also relaxed the PE (1 µM) induce contractions at 5 mg/mL with EC_{50} 15.53 mg/mL (95% CI: 7.965 to 62.27 mg/mL; n = 5) and 3 mg/mL with EC_{50} 4.2 mg/mL (95% CI: 2.991 to 6.670 mg/mL; n = 5), respectively. Similarly, verapamil also caused relaxation of K^+(80 mM) and PE (1 µM) induced contractions with respective EC_{50} 1.054 µM (95% CI: 0.45–5.68) and 0.764 µM (95% CI: 0.33–68.8). The aqueous fraction of (Hh.Aq) partially exerted relaxation of K^+(80 mM) and PE (1 µM) induced contractions on aortic preparation (Figure 5).

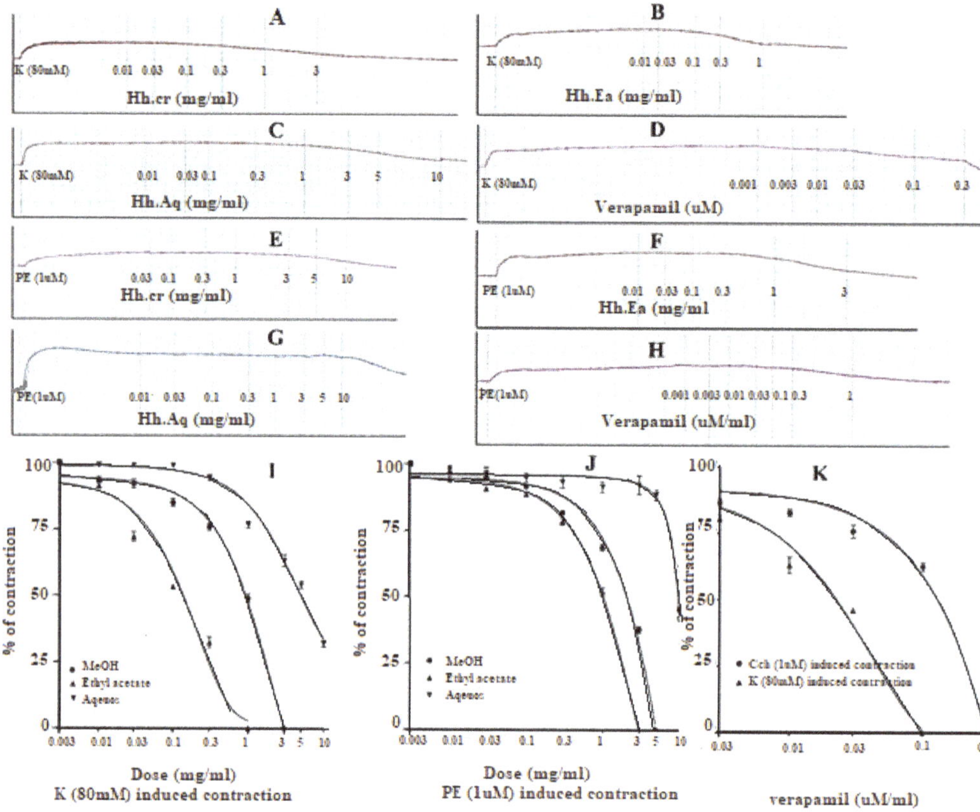

Figure 5. Effect of (**A**) crude extract (Hh.Cr), (**B**) Ethyl acetate fraction (Ea.Hh), (**C**) Aqueous fraction (Ea.Aq), and (**D**) verapamil on K^+ induced contraction and effect of (**E**) crude extract (Hh.Cr), (**F**) Ethyl acetate fraction (Ea.Hh), (**G**) Aqueous fraction (Ea.Aq), and (**H**) verapamil on PE 1 μM Induced Contraction on rabbit aorta preparations. Effect of crude extract (Hh.Cr) Ethyl acetate fraction (Ea.Hh) and Aqueous fraction (Ea.Aq) on (**I**) K^+ (80 mM) Induced Contraction and (**J**) PE 1 μM Induced Contraction on aortic jejunum preparations. (**K**) Effect of verapamil on PE 1 μM and K^+ Induced Contraction on rabbit tracheal preparations. Values are expressed as mean ± SEM.

2.6. Effect on Atria Preparations

The crude extract (Hh.Cr) and its ethyl acetate (Hh.Ea) fraction caused the negative chronotropic effect (i.e., decrease in heart rate) and negative inotropic effect (i.e., force of contraction) on atrium preparation [12]. Hh.Cr and its Hh.Ea fraction showed negative inotropic effect within concentration range 0.01–5.0 mg/mL with EC_{50} 0. 9 mg/mL (95% CI: 0.375–1.356 mg/mL; n = 3) and 0.01–3.0 mg/mL with EC_{50} 0.7 mg/mL (95% CI: 0.265–0.586 mg/mL; n = 3), respectively. Hh.Cr and its Hh.Ea fraction showed the negative chronotropic effect within concentration range 0.01–5.0 mg/mL with the EC_{50} 0.5 mg/mL (95% CI: 0.406–0.680 mg/mL; n = 3) and 0.01–3.0 mg/mL with the EC_{50} value calculated to be 0.4 mg/mL (95% CI: 0.106–0.050 mg/mL; n = 3), respectively. Similarly, verapamil also showed negative inotropic and chronotropic effect with concentration range 0.01–1.0 mg/mL with EC_{50} value of 0.053 μM (95% CI: 0.034–0.084 μM; n = 3) and 0.01–0.3 mg/mL with EC_{50} 0.037 μM (95% CI: 0.024–0.045 μM; n = 3). The aqueous fraction (Hh.Aq) exerted partially negative inotropic and chronotropic effect on atrium preparation with in concentration range 3–10 mg/mL with EC_{50} 1.02 mg/mL (95% CI:

0.485–0.856 mg/mL; n = 3) and 3–10 mg/mL with 1.14 mg/mL (95% CI: 0.575–1.756 mg/mL; n = 3), respectively (Figure 6).

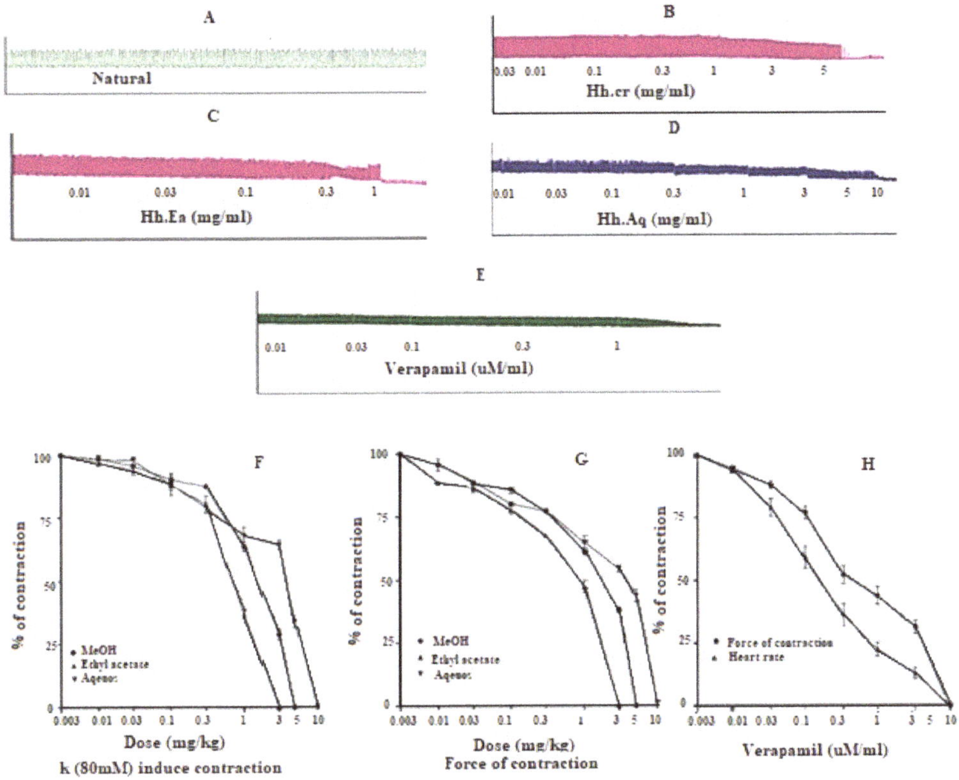

Figure 6. (**A**) Spontaneous contraction. Effect of (**B**) crude extract (Hh.Cr), (**C**) Ethyl acetate fraction (Ea.Hh), and (**D**) Aqueous fraction (Ea.Aq) on spontaneous contraction rabbit paired atrium preparations. (**E**) Effect of verapamil on spontaneous contraction rabbit paired atrium preparations. Effect of crude extract (Hh.Cr) Ethyl acetate fraction (Ea.Hh) and Aqueous fraction (Ea.Aq) on (**F**) K^+(80mM)-induced contraction. (**G**) force of contraction. (**H**) Effect of verapamil on the force of contraction and heart rate on rabbit atrium preparations. Values are expressed as mean ± SEM.

2.7. Antiperistalsis Activity

The crude extract (Hh.Cr) showed a significant antiperistalsis response in mice with less distance traveled by charcoal meal as compared to control (33 ± 2.3%). The group was treated with 400 mg/kg of Hh.Cr and CCh (3 mg/kg) and peristaltic movements were significantly decreased by 1.2 ± 0.37 and 8.6 ± 1.8, respectively (Figure 7).

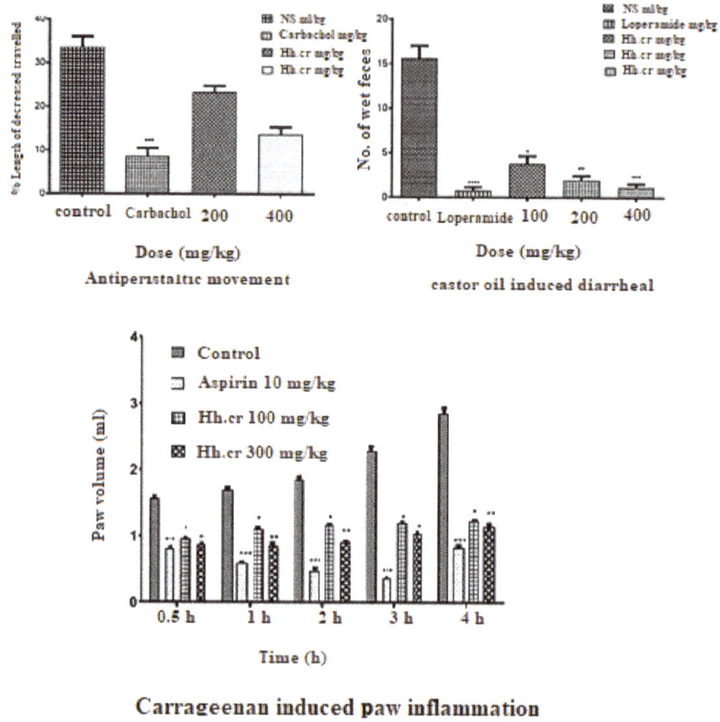

Figure 7. GI Charcoal meal transit (antiperistalsis) activity, castor oil-induced diarrhea activity, and carrageenan induce inflammation. Values are expressed as Mean ± SEM, and data was analyzed One way ANOVA or Two way ANOVA; * $p < 0.05$, ** $p < 0.005$, *** $p < 0.0005$ and **** $p < 0.0001$.

2.8. Antidiarrheal Activity

The crude extract (Hh.Cr) showed a significant antidiarrheal response in rats with fewer wet fecal masses than control (15.60 ± 1.4). The group was treated with 400 mg/kg of Hh.Cr and loperamide (3 mg/kg) showed highly significant anti-diarrheal effect 1.2 ± 0.37 and 0.8 ± 0.37, respectively (Figure 7).

2.9. Anti-Inflammatory Activity

The crude extract (Hh.Cr) showed a significant anti-inflammatory response in rats with inhibition of edematous volume of hind paw as compared to control (2.93 ± 0.2 mL) at maximum duration. The Hh.Cr inhibited the paw edema at 100 mg/kg as 0.96 ± 0.08 mL, 1.10 ± 0.01 mL, 1.16 ± 0.01 mL, 1.12 ± 0.015 mL, and 1.23± 0.01 mL at 0, 1, 2, 3, and 4 h duration, respectively, whereas at dose 300 mg/kg, it showed maximum inhibition, i.e., 0.87 ± 0.01 mL, 0.86 ± 0.01 mL, 0.86 ± 0.03 mL, 0.91 ± 0.01 mL, 1.04 ± 0.001 mL, and 1.15 ± 0.02 mL at 0, 1, 2, 3, and 4 h duration, respectively. Hh.Cr inhibited the edematous volume as similar to the aspirin, i.e., 1.05 ± 0.02 mL (Figure 7).

2.10. In Silico Studies

The docking calculations are beneficial to predict ligand pose within the binding site of the target protein. The involvement of physical energies terms (i.e., solvation energy) with suitable force field make docking calculation of compounds more acceptable with accuracy (Table 2) [12–14].

Table 2. Binding energies (kcal/mol) of compounds with Muscarinic-3 (MM3, PDB ID: 4U14), Cyclooxygenase-2 (COX-2, PDB ID:5IKQ) and lipoxygenase 5 (LOX-5, PDB ID: 6N2W) receptors calculated by Prime MMGBSA.

Name (PubChem ID)	Docking Score	$\Delta G_{Binding}$	Log K_i (μMolar)	$\Delta G_{Coulomb}$	$\Delta G_{Covalent}$	ΔG_{Hbond}	$\Delta G_{Lipophilic}$	$\Delta G_{Solv\ GB}$	ΔG_{vdW}	Residue-Ligand Interactions with Distance (Å)	
										Hydrogen Bonds	Hydrophobic Bonds
Arctiin (100528)	**−11.63**	**−60.79**	**−23.17**	**−16.29**	**10.56**	**−1.22**	**−42.47**	**47.40**	−55.32	Asn513 (2.46), Leu225 (1.77), **C-H Bond:** Thr231 (2.58), Tyr529 (2.99), **π-Donor Hydrogen Bond:** Trp525 (2.71), Trp525 (2.66)	**π-Sulfur Bond:** Cys532 (5.42), **π-π Stacked Bond:** Trp503 (4.56), **π-π T shaped Bond:** Tyr148 (5.21), **Alkyl Bond:** Ile222 (4.69), Leu225 (4.41), Cys532 (3.20), **π-Alkyl Bond:** Tyr148 (3.64), Tyr506 (3.99), Tyr529 (4.91), Tyr529 (3.98), Leu225 (4.98)
Arctigenin (64981)	−9.72	−46.56	−16.99	−16.92	6.57	−0.68	−30.61	31.10	−34.45	Ala238 (1.64) **C-H Bond:** Thr234 (3.03), Ile222 (2.77), Leu225 (2.77), Leu225 (2.65), Tyr148 (2.40)	**π-π T shaped Bond:** Trp503 (5.58), Trp525 (5.37), **π-Alkyl Bond:** Tyr148 (4.40), Trp199 (4.19), Trp199 (4.18), Phe221 (4.72), Trp525 (3.99), Trp525 (4.05), Leu225 (5.49), Ala238 (4.30)
Catechin (9064)	−7.59	−52.22	−19.45	−29.22	2.89	−2.92	−13.93	27.23	−34.12	Tyr148 (2.05), Ile222 (2.39), Ile222 (3.03), Ser226 (1.92), Ser226 (1.80)	**π-π T shaped Bond:** Tyr506 (5.74)
Chlorojanerin (182408)	−7.13	−43.68	−15.74	−20.79	2.92	−1.98	−19.31	33.51	−38.02	Tyr127 (1.87), Tyr148 (2.30), Asn513 (3.02) Asn513 (3.03), Asn526 (2.03), Ser226 (2.60), **C-H Bond:** Leu225 (2.63), Leu225 (2.55), Ser226 (2.55)	**Alkyl Bond:** Lys522 (5.21), Lys522 (5.47), **π-Alkyl Bond:** Phe124 (4.85), Trp525 (4.93), Trp525 (5.15), Trp525 (3.59), Trp525 (4.43)
Cynaropicrin (119093)	−6.76	−48.69	−17.92	−19.54	1.54	−1.63	−18.37	26.11	−36.81	Tyr148 (3.00), Ile222 (2.49), Asn526 (1.77), Leu225 (1.92), Thr231 (2.64)	**Alkyl Bond:** Lys522 (5.30), **π-Alkyl Bond:** Phe124 (5.48), Tyr127 (4.81), Trp525 (3.49), Trp525 (4.36)

Table 2. *Cont.*

Name (PubChem ID)	Docking Score	$\Delta G_{Binding}$	Log K_i (μMolar)	$\Delta G_{Coulomb}$	$\Delta G_{Covalent}$	ΔG_{Hbond}	$\Delta G_{Lipophilic}$	$\Delta G_{Solv\ GB}$	ΔG_{vdW}	Residue-Ligand Interactions with Distance (Å)	
										Hydrogen Bonds	Hydrophobic Bonds
Cyclooxygenase-2 (COX-2, PDB ID:5IKQ)											
Arctiin (100528)	−8.49	−41.01	−14.58	−21.01	10.60	−2.56	−23.77	33.40	−36.96	Lys83 (1.88), Ser12 (3.00), Ser120 (1.73), Pro84 (1.67), **C-H Bond:** Ser120 (2.66)	**π-π T shaped Bond:** Tyr11 (5.32), **Alkyl Bond:** Ala112 (3.50), Val89 (4.95), Leu93 (4.90), Val117 (4.37), Leu109 (4.79), Ile113 (5.11), **π-Alkyl Bond:** Tyr116 (4.01), Val89 (3.76), Le113 (4.71)
Arctigenin (64981)	−7.37	−27.91	−8.89	−5.55	18.02	0.00	−31.09	26.83	−35.44	**C-H Bond:** Ala528 (2.91), Ser120 (2.62), Ser531 (2.79)	**π-σ Bond:** Val117 (2.48), **Alkyl Bond:** Arg121 (4.83), Val350 (4.80), Leu353 (5.46), Val89 (4.59), Leu93 (5.01), **π-Alkyl Bond:** Val350 (5.15), Leu353 (4.99), Val524 (4.56) Ala528 (4.24)
Cynaropicrin (119093)	−4.28	−35.97	−12.39	−11.96	3.10	−1.10	−18.58	17.86	−25.30	Arg121(1.81), Arg121 (2.46), **C-H Bond:** Val117 (2.51)	**Alkyl Bond:** Pro84 (5.14), Val89 (5.04), Val89 (4.09), Pro84 (4.81), Val89 (4.45), Ile92 (5.00), Leu93 (3.76), **π-Alkyl Bond:** Tyr116 (5.22)
Catechin (9064)	−2.84	−11.87	−1.93	2.05	4.15	−0.61	−11.82	15.82	−18.71	Arg121 (2.76), Tyr116 (2.79), **C-H Bond:** Pro84 (2.55), Tyr116 (2.29)	**π-π T shaped Bond:** Tyr116 (5.65), Tyr116 (4.80), **Alkyl Bond:** Val89 (4.00), **π-Alkyl Bond:** Tyr116 (5.27), Val89 (4.80), Pro84 (5.20)
Lipoxygenase 5 (LOX-5, PDB ID: 6N2W)											
Arctiin (100528)	−5.76	−30.76	−10.13	−14.19	6.04	−2.11	−16.17	44.56	−46.53	His372 (2.55), Glu417 (1.89), **C-H Bond:** Glu417 (3.00), Gln413 (2.62)	**Electrostatic π-Anion Bond:** Ile673 (4.46) **π-π Stacked Bond:** His372 (4.40), **Alkyl Bond:** Ala410 (3.62), Leu368 (4.59), Leu368 (4.40), **π-Alkyl Bond:** His367 (3.52), His372 (4.83), His372 (3.83), Ile406 (5.49), Ala410 (4.09)

Table 2. *Cont.*

Name (PubChem ID)	Docking Score	$\Delta G_{Binding}$	Log K_i (µMolar)	$\Delta G_{Coulomb}$	$\Delta G_{Covalent}$	ΔG_{Hbond}	$\Delta G_{Lipophilic}$	$\Delta G_{Solv\ GB}$	ΔG_{vdW}	Residue-Ligand Interactions with Distance (Å)	
										Hydrogen Bonds	Hydrophobic Bonds
Catechin (9064)	−4.95	−30.81	−10.15	−22.76	5.66	−2.38	−15.10	39.78	−32.68	Arg596 (2.34), His600 (1.80), **π-Donor Hydrogen Bond:** His372 (3.10)	**π-π Stacked Bond:** His367 (4.84), **π-π T shaped Bond:** His372 (5.54), Trp599 (5.02), **Alkyl Bond:** Leu607 (5.03), **π-Alkyl Bond:** Leu607 (5.21), Ala603 (4.88)
Arctigenin (64981)	−4.84	−42.94	−15.42	−28.08	3.66	−3.06	−18.99	34.43	−29.61	Arg596 (2.57), Arg596 (1.88), His600 (1.82)	**Electrostatic π-Cation Bond:** Arg596 (3.17), **Alkyl Bond:** Ala410 (3.75), Ala426 (3.63), **π-Alkyl Bond:** His367 (4.32), Trp599 (4.00), Leu607 (5.10), Ala426 (3.91)
Cynaropicrin (119093)	−3.45	−14.54	−3.09	−13.54	2.95	−0.81	−17.02	46.96	−33.09	His367 (2.76), Ile673 (1.84), **C-H Bond:** Ala410 (2.98)	**Alkyl Bond:** Ala603 (4.92), Ala603 (3.87), Leu607, (5.09), Leu607 (4.33)
Chlorojanerin (182408)	−3.30	−32.60	−10.93	−3.30	0.38	−0.62	−13.65	19.69	−35.10	Thr427 (2.78), Arg596 (1.77), His600 (2.15), **C-H Bond:** His367 (2.97), His600 (2.56), Pro569 (2.73)	**π-π T shaped Bond:** Trp599 (5.64), **Alkyl Bond:** Ala603 (3.39), Val604 (4.44), **π-Alkyl Bond:** His360 (5.21), His432 (4.65), Trp599 (4.91) His600 (4.18)

$\Delta G_{Binding}$: Binding free energy, Log Ki: Logarithmic of Inhibition Constant (K_i), $\Delta G_{Coulomb}$: Coulomb binding energy, $\Delta G_{Covalent}$: Covalent binding energy ΔG_{Hbond}: Hydrogen bonding energy, $\Delta G_{Lipophilic}$: Lipophilic binding energy, $\Delta G_{Solv\ GB}$: Generalized born electrostatic solvation energy ΔG_{vdW}: Van der Waals forces energy, and C-H Bond: Carbon–Hydrogen Bond. These all contribute to Binding free energy ($\Delta G_{Binding}$).

Molecular docking for Muscarinic M3 receptor: The selected compounds were studied against muscarinic M3 (MM3, PDB ID: 4U14)) for antispasmodic activity (Table 2, Figure 8). Arctiin was predicted with the lowest binding energy (ΔG_{bind}: −60.79 kcal/mol) with hydrophobic energies ΔG_{vdW} (−55.32 kcal/mol) and ΔG_{Lipo} (−42.47 kcal/mol) major contributors to the ligand binding energy. It formed the two π-donor hydrogen interaction with residue Trp525 and hydrophobic interactions (π-π Stacked Bond: Trp503; π-π T shaped Bond: Tyr148) within the pocket of MM3. Besides these, it also formed the π-Sulfur interaction with residue Cys532. Arctigenin second to arctiin also found potent with have ligand binding energy (ΔG_{bind}: −46.56 kcal/mol) mainly contributed with hydrogen bond interaction (ΔG_{Hbond}: −0.68 kcal/mol) and hydrophobic interaction (ΔG_{vdW}: −34.45 kcal/mol and ΔG_{Lipo}: −30.61 kcal/mol). It formed hydrophobic π-π T-shaped interaction with residue Trp503 and Trp525 within the protein cleft. Moreover, catechin has the lowest binding energy (ΔG_{bind}: 52.22 kcal/mol) with ΔG_{vdW} (−34.12 kcal/mol) and ΔG_{Lipo} (−13.93 kcal/mol) and formed π-π T shaped interaction with residue Tyr506. The ranking orders of ligands with COX-2 is given below: arctiin > arctigenin > catechin > chlorojanerin > cynaropicrin.

Molecular docking for cyclooxygenase-2 enzyme: The selected compounds were studied against cyclooxygenase-2 enzyme (COX-2, PDB ID:5IKQ) for anti-inflammatory activity (Table 2, Figure 8). Arctiin was predicted with the lowest binding energy (ΔG_{bind}: −41.01 kcal/mol) among the selected compounds. As mentioned earlier, Van der Waals (ΔG_{vdW}) and lipophilic interactions (ΔG_{Lipo}) are significant contributors to the ligand binding energy. It was observed that the binding energies value of ΔG_{vdW} was −36.96 kcal/mol, and ΔG_{Lipo} was −23.77 kcal/mol, whereas hydrogen bond (ΔG_{Hbond}) energy contribution was −2.56 kcal/mol. It also formed hydrophobic interactions π-π T shaped interaction with Tyr11. Arctigenin second to arctiin in docking score was found with potential hydrophobic interactions within hydrophobic clefts of COX-2. The ligand binding energy of arctigenin (ΔG_{bind}: −27.91 kcal/mol) was driven mainly by these hydrophobic interaction energies; Δ G_{vdW} (−35.44 kcal/mol) and ΔG_{Lipo} (−31.09 kcal/mol) and formed hydrophobic π–σ interaction with Val117. The ranking orders of ligands with COX-2 are given below: arctiin > arctigenin > cynaropicrin > catechin.

Molecular docking for lipoxygenase-5 enzyme: The selected compounds were studied against lipoxygenase-2 enzyme (LOX-5, PDB ID: 6N2W) for anti-inflammatory activity (Table 2, Figure 8). The contribution of hydrophobic interactions in ligand binding energy was more abundant within pockets of LOX-5. Arctigenin has higher ligand binding energy (ΔG_{bind}: −42.94 kcal/mol) but ranks third in the docking score. The ligand binding energy contributed with hydrogen bond interaction (ΔG_{Hbond}: −3.06 kcal/mol) and hydrophobic interaction (ΔG_{vdW}: −29.61 kcal/mol and ΔG_{Lipo}: −18.99 kcal/mol). It also formed hydrophobic interactions (π-π Stacked Bond: His372) within the pocket of LOX-5. Arctiin ranked at first in position docking score with ligand binding energy (ΔG_{bind}: −30.76 kcal/mol) which mainly contributed from hydrophobic interaction energies ΔG_{vdW} (−46.53 kcal/mol) and ΔG_{Lipo} (−16.17 kcal/mol), whereas hydrogen bond (ΔG_{Hbond}) energy contribution was −2.11 kcal/mol. Besides hydrophobic interaction, arctigenin and arctiin also formed electrostatic charge interaction with residue Arg596 and Ile673 within the cleft of COX-2, respectively. Catechin have ligand binding energy (ΔG_{bind}: −30.81 kcal/mol) mainly contributed with hydrogen bond interaction (ΔG_{Hbond}: −2.38 kcal/mol) and hydrophobic interaction (ΔG_{vdW}: −32.68 kcal/mol and ΔG_{Lipo}: −15.10 kcal/mol). Catechin formed the π–donor hydrogen interaction with residue His372 and hydrophobic interactions (π-π Stacked Bond: His367; π-π T shaped Bond: His372, Trp599) within the pocket of LOX-5. The ranking order of ligands with COX-2 is given below: arctiin > catechin > arctigenin > cynaropicrin > chlorojanerin

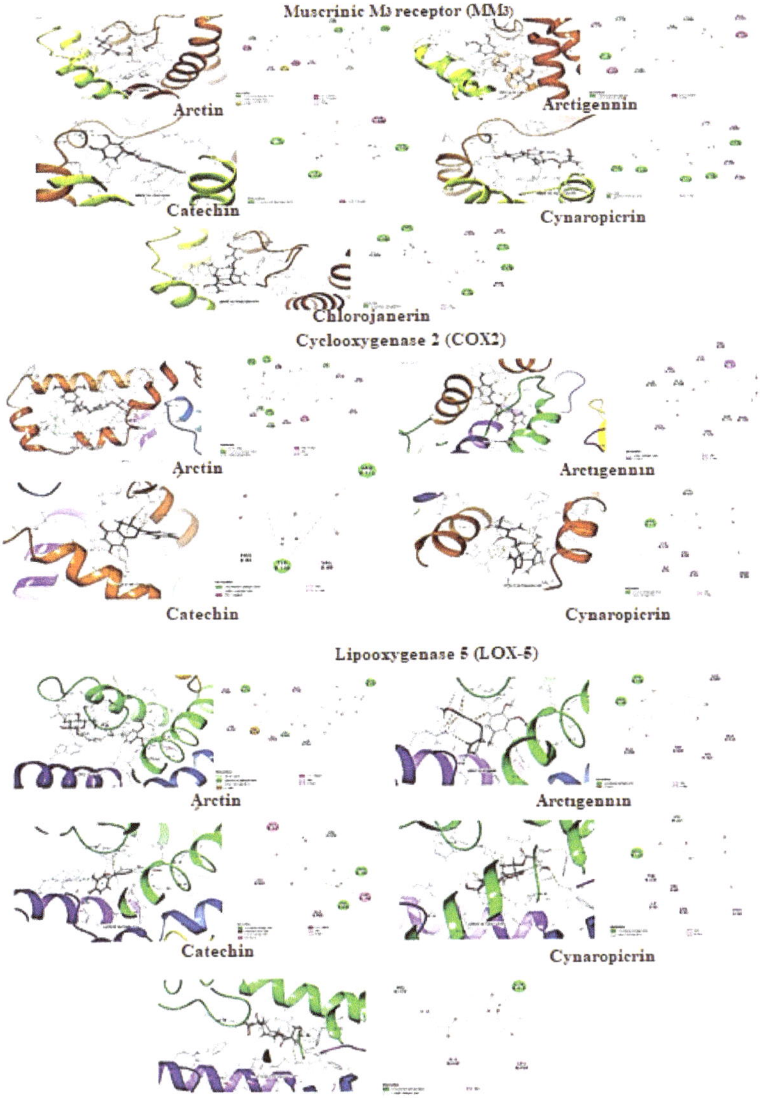

Figure 8. Molecular docking of selected compounds against muscarinic receptor, cyclooxygenase-2, and lipoxygenase 5.

3. Discussion

Himalaiella heteromalla has a potential pharmacological role in the management of various diseases. This research was employed to investigate its pharmacological characteristics. The presence of alkaloids, glycosides, triterpenoids, flavonoids, saponins, sesquiterpene, which play a vital role in the pharmacological potential of *Himalaiella heteromalla*. The HPLC results indicate the presence of gallic acid, catechin, HB acid, and vanillin acid. Gallic acid (3,4,5trihydroxybenzoic acid), a natural polyphenol product, has anti-oxidant, anti-inflammatory, antimicrobial, and radical scavenging activities. Gallic acid is used as a spasmolytic effect on smooth muscle isolated jejunum tissues and trachea by calcium channel blocking activity [15]. Gallic acid is used as an antispasmodic in diarrhea [16]. Gallic acid possesses an anti-inflammatory effect [17]. Catechin abundant flavonoid present in plants, it reported for several gastrointestinal, respiratory, and inflammatory disorders [18–20]. Vanilla acid and HB acid are polyphenolics used for gastrointestinal, respiratory, and cardiovascular disorders by spasmolytic effects on isolated tissues of the jejunum, trachea, and aorta [20]. The water content in *H. heteromalla* play a vital role in the biological activities, so it is more important to measure the water content in *H. heteromalla* therefore, infra red radiation can be used to measure water content determination [21].

Himalaiella heteromalla crude extract (Hh.Cr) was studied on isolated jejunum preparations to elaborate the mechanism of *H. heteromalla* in gastrointestinal diseases. It is reported that jejunum preparations have rhythmic contractions due to the influx of calcium ions and potassium ions through their respective ions channels. *H. heteromalla* crude extract and its fraction ethyl acetate exerted spasmolytic response in dose concentration when exposed to spontaneous contraction of jejunum preparations [22]. Thus, *H. heteromalla* crude extract and its fraction ethyl acetate showed the antispasmodic response by suppressing rhythmic contractions in jejunum preparations. These results indicate that *H. heteromalla* crude extract and its fraction ethyl acetate decrease or blockade the cytoplasmic free Ca^{++} ions through the blockade of voltage-dependent calcium ion channels. As a result, activation of calmodulin and other contractile proteins, i.e., actin and myosin, does not occur [23]. *H. heteromalla* crude extract and its fraction ethyl acetate and aqueous were exposed to K^+ (80 mM)-induced contractions on jejunum preparations, *H. heteromalla* crude extract and its fraction ethyl acetate relaxed the K^+ (80 mM)-induced contractions in dose concentration manner in a tissue organ bath. It was previously reported that K^+ (80 mM) induces contractions to cause cell depolarization by the influx of calcium ions into the cell through the voltage-gated L-type calcium ion channel [24]. Similar to verapamil, any substance inhibited K^+ (80 mM)-induced contractions were considered calcium channel blockers (CCB). Thus, *H. heteromalla* crude extract and its fraction ethyl acetate blockade the calcium influx into the cell by alternating or binding with voltage-dependent calcium channels. Furthermore, calcium concentration–response curves (CRCs) were constructed on pretreated Hh.Cr jejunum preparations to confirm the calcium channel blockade activity of Hh.Cr in a tissue organ bath. The results showed that partial blockade with the rightward parallels dose–response curves at low doses while completely blocking the dose–response curves at 0.3 mg/m. Thus, *Himalaiella heteromalla* exhibited a strong calcium antagonistic effect [25].

To evaluate another possible mechanism of *H. heteromalla* crude extract on the gastrointestinal tract, Hh.Cr was studied in antiperistalsis and antidiarrheal in vivo models. *H. heteromalla* crude extract showed the antispasmodic response by inhibiting the traveling of charcoal meal in antiperistalsis activity. *H. heteromalla* crude extract also inhibited diarrheal response in castor oil-induced diarrhea. It decreased the wet feces by inhibiting the electrolyte and water imbalance that may cause diarrhea in rats [26].

H. heteromalla crude extract and its fractions ethyl acetate and aqueous were tested for possible bronchodilator activity against CCh (1 μM) and K^+ (80 mM)-induced contractions on tracheal preparations. The results showed that *H. heteromalla* crude extract and its fractions ethyl acetate exhibited relaxant response against CCh (1 μM) and K^+ (80 mM)- induced contractions, but a partial relaxant effect was observed by the aqueous fraction. However, EC_{50} of *H. heteromalla* crude extract and its fractions ethyl acetate against K^+ (80 mM)-induced contractions that were more minor than CCh-induced contractions, similar to that of verapamil. CCh is a cholinergic agonist which causes smooth muscle contraction through activation of muscarinic receptors. Hence, *H. heteromalla* crude extract and its fractions ethyl acetate showed bronchodilator response was found due to Ca^{++} ion channel and muscarinic receptor blockade. Nowadays, Ca^{++} channel blockers and muscarinic antagonists are used to treat the relief from respiratory diseases such as asthma [27,28].

H. heteromalla crude extract and its fractions ethyl acetate and aqueous were tested for possible vasorelaxant activity against PE (1 μM) and K^+ (80 mM)-induced contractions on aortic preparations. The results showed that *H. heteromalla* crude extract and its fractions ethyl acetate exhibited a relaxant response against PE (1 μM) and K^+ (80 mM)-induced contractions, similar to verapamil. *H. heteromalla* aqueous fraction partially relaxed the PE and K^+ (80 mM) induced contractions. Relaxation of the PE (PE) and K^+ (80 mM)-induced contractions indicates a blockade of intracellular Ca^{++} influx by blocking Ca^{++} channels. Ca^{++} channel blockers are essential drugs used clinically to manage angina and hypertension [29,30].

H. heteromalla crude extract and its fractions ethyl acetate and aqueous were tested on paired atrium for possible effects on force and rate of atrial contractions. *H. heteromalla* crude extract and its fractions ethyl acetate and aqueous showed cardio suppressant response via blocking calcium channels, hence Hh.Cr and its fractions were found with adverse inotropic and chronotropic effects on the paired atrium [31].

Himalaiella heteromalla crude extract was tested for anti-inflammatory activity. It was found that Hh.Cr blocked the release of inflammatory mediators in rat paw edema and other models. It is reported that carrageenan acetic acid and formalin release inflammatory mediators such as bradykinin, histamine, TNF, IL-1b, IL-6, PEG2, and TNF were blocked by crude extract of crude extract *Himalaiella heteromalla*. The reduction in inflammatory mediators by carrageenan, inducing the rat's paw edema model to show that Hh.Cr inhibits factors that cause inflammation and swelling. On the other side, pain sensation is a significant indicator in the inflammation process, which Hh blocked Cr, so that it exhibited analgesic activity. This anti-inflammatory result was compared with standard drug analgesic aspirin, reducing all models' inflammation and pain sensations. The comparative results in between aspirin and Hh.Cr showed that Hh.Cr exhibited thde same potential as aspirin to reduce the pain and inflammation via blockading inflammatory mediators [32].

Molecular docking is a helpful tool to predict the possible mechanism of actions of the selected compounds of various pharmacological studies—the present study correlated and defined antispasmodic and anti-inflammatory activities of *Himalaiella heteromalla*. The five compounds of *H. heteromalla* were studied for cyclooxygenase 2, lipoxygenase 5, and muscarinic M3 receptor. The docking calculations of these compounds indicate the presence of antispasmodic and anti-inflammatory activities, which were previously proven in experimental studies. Arctiin and arctigenin were more potent compounds responsible for these activities. These results conclude that these compounds interact with cyclooxygenase 2, lipoxygenase 5, and muscarinic M3 receptor to exert the activity. As mentioned earlier, Gao et al. [4] reported that arctigenin and arctiin have anti-inflammatory and vasodilation properties and help treat acute lung injury, local edema, brain trauma, and colitis. These studies support the potent results of arctigenin and arctiin in silico studies against major inflammatory proteins COX-2 and LOX-5.

The pretreatment of various chemical, physical, physicochemical, and biological methods have been suggested to improve enzymatic hydrolysis; these techniques can be improved by the activities of extract [33–35]. In addition, the pretreatment and pyrolysis process require a high amount of external heat for (1) drying the washed biomass, (2) biomass torrefaction, and (3) biomass pyrolysis. Biomass drying and pyrolysis require a high amount of external heat for drying and torrefaction [36,37]. The plant also contains potassium (K), calcium (Ca), sodium (Na), and magnesium (Mg), which will significantly affect the behaviors of extract activity. It is essential to remove the metals by adopting different methods [38].

4. Materials and Methods

4.1. Extract Preparation

Himalaiella heteromalla (D.Don) Raab-Straube was collected from hilly areas of Islamabad and identified by Dr. Zafarullah Zafar, taxonomist, Institute of Pure and Applied Biology, and submitted with voucher no: http://www.theplantlist.org/tpl1.1/record/gcc-138921 dated 18 June 2018. Plant material was ground through a herbal grinder for coarse powder, then powder (1 kg) was macerated in methanol aqueous (70:30) for maceration in an amber color glass bottle for three days at room temperature and periodically shaken 3–4 times a day. The solvent was filtered to remove plant debris with muslin cloth and Whattman-1 filter paper. This procedure was replicated thrice, and the filtrate obtained by all steps was combined and processed in a rotary evaporator (BUCHI) under reduced pressure at 36 ± 2 °C to obtain a brownish colored semi-solid (Hh.Cr) and stored at −20 °C in an airtight jar with a percentage yield of 12%. The Hh.Cr (20 g) was subjected to solvent–solvent extraction with ethyl acetate and distilled water to produce an ethyl-acetate fraction (Hh.Ea) and aqueous fraction (Hh.Aq) with approximately 5.5% and 40% yield, respectively. *H. heteromalla* crude extract (Hh.Cr) and its fractions were moderately soluble in aqueous. All dilutions were prepared fresh on the day of the experiment.

4.2. Animal Housing

Both sexes of albino mice (weight: 20–30 g), rats (weight: 150–200 g), and rabbits (weight: 1–1.8 kg) were used in this study and kept under controlled housing conditions with a temperature of 23 ± 3 °C in the animal house of the Faculty of Pharmacy, Bahauddin Zakariya University, Multan. Before the experiment, animals were deprived of food overnight but had free access to water. For in vitro experimentation, rabbits were sacrificed following a blow, while mice and rats were killed by cervical dislocation. All the experimentations were performed under rules specified by the Institute of Laboratory Animal Resources, Commission on Life Sciences (NRC, 1996) endorsed by the Ethical Committee of Bahauddin Zakariya University, Multan.

4.3. Chemicals

All the chemicals used in this study have high purity with research-grade quality. Acetylcholine (Ach), aspirin, carbamylcholine chloride HCl, Carbachol (CCh), verapamil HCl, phenylephrine (PE) were purchased from Sigma Chemical Company, St. Louis, MO, USA. While Potassium dihydrogen phosphate, magnesium chloride, sodium bicarbonate, sodium chloride, magnesium sulfate, sodium dihydrogen phosphate, calcium chloride, potassium chloride, ethylenediaminetetraacetic acid (EDTA), glucose were purchased from Merck, Dermstadat Germany. Furthermore, loperamide, and dicyclomine were supplied by Sigma Chemical company, St. Louis, MO, USA).

4.4. Qualitative Phytochemical Detection

The qualitative phytochemical investigation of *H. heteromalla* was performed to identify alkaloids, glycosides, anthraquinones, terpenes, saponins, flavonoids, and phenols.

4.5. HPLC Separation of Phenolic Acids and Flavonoids

The phenolic acids and flavonoids components in *Himalaiella heteromalla* were quantified by developing a binary gradient solvent system to run in Chromera HPLC system (Perkin Elmer, Houston, TX, USA) consisting of Felexer Binary Liquid chromatography (LC) pump coupled with UV/Vis LC Detector (Shelton, CT, USA) which was operated with the help of a software. HPLC system consisted of a C-18 column (250 × 4.6 mm internal diameter) with a thickness of 5 μM film. The mobile phase consisted of solvent A (methanol (30): acetonitrile (70)) and solvent B (0.5% glacial acetic acid in double-distilled water), mobile phase run at flow rate 0.08 mL/min, and data was recorded at 275 nm of UV spectra. The peaks and retention times of phenolic acids and flavonoids of *H. heteromalla* were matched with external standards to quantify the components [12]. The resolution and separation factor was used to determine HPLC separation efficiency.

4.6. In Vitro Experiments

The physiological response of tissues was recorded with isotonic and force-displacement isometric transducers amplified with acquisition system Power Lab (AD Instruments, Bella Vista, NSW, Australia) coupled with a computer having Lab chart Pro. The effect was taken as percent change on the part of test substance recorded instantly preceding a dose of test substance [22].

4.6.1. Isolated Rabbit Jejunum Preparation

The jejunum was dissected from a rabbit; the adhesive fatty tissues were carefully removed, and then ~2 cm long piece of jejunum was prepared. This tissue was hung in an organ bath containing Tyrode's solution with a continuous supply of carbogen (95% O_2 + 5% CO_2) at 37 °C and equilibrated for 30 min. Acetylcholine (1 µM) was added to spontaneous rhythmic contractions of jejunum for control response and washed it. The Hh.Cr was added cumulatively for antispasmodic effect. The spontaneous contractions jejunum preparation was exposed to K^+ (80 mM) induced contraction for estimation of CCB activity [39].

The extract was exposed to calcium concentration response curves (CRCs) for further confirmation. The jejunum preparation was stabilized in Tyrode's solution, subsequently replaced with calcium-free Tyrode's solution with EDTA (0.1 mM) to remove calcium from tissue. Afterward, with an incubation duration of 40 min, two superimposable control calcium CRCs were constructed in an organ bath, then tissue was incubated with the plant exact for one h, and calcium CRCs were obtained and compared to control. The calcium CRCs were recorded in the presence of different concentrations of plant extract.

4.6.2. Isolated Rabbit Tracheal Preparations

The trachea was dissected from a rabbit for bronchodilator activity, and the 2 mm tracheal ring tissue was prepared. A longitudinal incision was made opposite the smooth muscle layer to form a strip. This tracheal preparation was hung in an organ bath containing Krebs's solution with a continuous supply of carbogen at 37 °C. Preload tension (1 g) was applied and allowed to equilibrate for 60 min prior to the dose of any drug. The tracheal preparation was exposed to CCh (1 µM), and K^+ (80 mM) induced contraction for bronchodilator activity in a cumulative manner.

4.6.3. Isolated Rabbit Paired Atria Preparations:

The heart was dissected from a rabbit for cardiac activity, and the ventricles were carefully removed to isolate paired atria. This atrium preparation was hung in an organ bath containing Krebs's solution with a continuous supply of carbogen at 34 °C. Then, 1 g preload tension was applied and allowed to equilibrate for 30 min prior to the dose of any drug. The isolated atrium preparation was exposed for possible cardiac effects in a cumulative fashion, and changes in rate and force of contractions were observed.

4.6.4. Isolated Rabbit Aorta Preparations

For vasorelaxant activity, the thoracic aorta was dissected from a rabbit, carefully removed the adhesive fatty tissues, and prepared 2–3 mm aortic rings. This aortic preparation was hung in an organ bath containing Krebs's solution with a continuous supply of carbogen. Preload tension (2 g) was applied and allowed to equilibrate for 60 min prior to the dose of any drug. The isolated aortic preparation was exposed to *H. heteromalla* for possible vasorelaxant effects in a cumulative manner. Further to define the possible mechanism, *H. heteromalla* was challenged to PE (1 µM), and K^+ (80 mM) induced contraction for the possible activity of *H. heteromalla* in a cumulative manner.

4.7. *In Vivo Activities*

4.7.1. Antiperistalsis Activity

Antiperistalsis activity was performed according to the method prescribed by Wahid et al. [12]. Mice (25) of either sex were divided into 5 groups, i.e., control (0.9% normal saline), standard drug (CCh10 mg/kg), and *H. heteromalla* doses (100, 200, and 400 mg/kg). After 15 min of administering the test or standard material orally, each animal received 0.3 mL of the charcoal meal (10% gum acacia, 20 starch, and 10% vegetable charcoal) in distilled water. Thirty minutes later, mice were killed, and the abdomen was incised to excise the whole small intestine. The distance from the pylorus region was measured to the front of the charcoal meal.

4.7.2. Antidiarrheal Activity

The antidiarrheal activity was performed according to the method prescribed by Wahid et al. [12] with modifications. Mice (20) of either sex were divided into five groups, i.e., negative control (0.9% normal saline), standard drug (loperamide 10 mg/kg), and *H. heteromalla* doses (100, 200, and 400 mg/kg). After 30 min of dose administration (p.o.), animals received the castor oil (10 mL/kg p.o) and were observed for six hours in cages with a white paper surface with adsorbent properties. The percent inhibition of wet fecal was calculated.

4.7.3. Carrageenan-Induced Rat's Hind Paw Edema Method

The anti-inflammatory activity was performed [32,40] on 25 rats of either sex divided into 5 groups, i.e., control (0.9% normal saline) standard drug (aspirin 10 mg/kg), and *H. heteromalla* doses (100, 200, and 400 mg/kg). After 30 min of dose administration (p.o), edema was induced by injecting 1% carrageenan in the right hind paw's sub-planter region and measuring the edema size at up to 4 h through a plethysmometer. The percentage of edema inhibition was calculated.

4.8. In Silico Studies

In silico studies were performed according to the method previously reported by Wahid et al. [12] and Sirous et al. [13].

Ligand Preparation: The 2D structures of HPLC quantified phytocompounds were retrieved from PubChem (https://pubchem.ncbi.nlm.nih.gov accessed date 20 March 2019) and treated in the LigPrep module of Maestro (Schrodinger suite 2015) to ionization, minimization, and optimization of ligands. The Epik tool of this module was used to generate the ionization state of ligands at cellular pH (7.4 ± 0.5) and applied the OPLS3e force field through the module for minimization and optimization of ligands that produce the lowest energy conformer of ligands.

Protein Preparation: For molecular docking, the highest resolution X-ray structures of proteins were downloaded from The Protein Databank (RCSB PDB) (https://www.rcsb.org accessed date 20 March 2019) and subjected to Protein preparation wizard of Maestro (Schrodinger suite 2015). This module processed the protein by adding hydrogen atoms to protein structure, removal of solvents (water) molecules, assigning bond orders, creating disulfide bonds, filling missing side chains and loops, and generating protonation state using Epik tool of protein structures for ligands at the cellular level pH (7.4 ± 0.5). After processing protein structures, these structures were optimized using PROPKA under pH 7.0, and the OPLS3e force field was utilized to perform restrained minimization for energy minimization and geometry optimization of protein structure.

Molecular Docking and Receptor grid generation: The active sites of protein structures for molecular docking were defined in the Receptor Grid Generation module of Maestro (Schrodinger suite 2015). A cubic grid box of each protein was defined with the help of a literature survey and with a selection of previously bonded ligands of proteins. The length of the grid box was adjusted to the length of 16 Å. The potential of nonpolar parts of the receptor was decreased to scaling factor 1.0 Å on Van der Waals radius of nonpolar atoms of protein having partial atomic charge cut-off 0.25 Å.

For molecular docking, the prepared ligands and protein structures were subjected to extra precision (XP) mode of Ligand Docking (Glide) module of Maestro (Schrodinger suite 2015) using pre-generated grid file for receptor. Additionally, 0.80 Å scaling factor was adjusted for Van der Waals radii with a partial charge cut-off of 0.15 Å. The docking results were subjected to the Prime MM-GBSA module to calculate the binding energies of ligands with protein structure using the VSGB solvation model with OPLS3e force field.

Inhibition Constant (K_i): The inhibition constant was determined from the binding free energy of ligand previously generated from Prime MM-GBSA, according to the following equation [12]:

$$\Delta G = -RT(\ln K_i) \text{ or } K_i = e^{(-\Delta G/RT)}$$

where ΔG is binding free energy of ligand, R is gas constant ($cal \cdot mol^{-1} \cdot K^{-1}$), and T is room temperature (298 Kelvin).

4.9. Statistical Analysis

The data were expressed as the mean ± standard error of the mean (S.E.M.) and median effective concentration (EC_{50}) with a 95% confidence interval (CI). One-way and two-way ANOVA tests were applied for in vivo experiments. All graphs and data were analyzed with the help of Graph pad prism software (San Diego, CA, USA).

5. Conclusions

Himalaiella heteromalla exhibited a more spasmolytic effect in ethyl acetate fraction and caused complete relaxation on isolated jejunum, trachea, aorta, and paired atria, supported with in silico studies. *H. heteromalla* proved various disease management-related activities. Further studies could be taken on *Himalaiella heteromalla* for drug discovery for the welfare of human beings.

Author Contributions: F.S. and S.M. planned the project and worked on statistical analysis of data and results interpretation; S.M. performed the experiments and F.U. worked on in-silico studies; R.A.M., C.C.M., N.B., N.U.-R. and M.R. drafted the manuscript. All authors have read and agreed to the published version of the manuscript.

Funding: This research was partially funded by FDI-0013.

Institutional Review Board Statement: Ethical approval was attained from Ethical Committee of Bahauddin Zakariya University, Multan (EC/25-MPHIL-S2018) dated 26 March 2018. Researchers agreed using the approved informed consent documented before their enrolment into study.

Informed Consent Statement: Not applicable.

Data Availability Statement: Not applicable.

Acknowledgments: The author(s) are very thankful Chairman Department of Pharmacology; Imran for Provision of animal house and lab related facilities.

Conflicts of Interest: Authors announce there are no conflict of interest concerning the publication of this research article.

References

1. Saklani, A.; Hegde, B.; Mishra, P.; Singh, R.; Mendon, M.; Chakrabarty, D.; Kamath, D.V.; Lobo, A.; Mishra, P.D.; Dagia, N.M.; et al. NF-κB dependent anti-inflammatory activity of chlorojanerin isolated from Saussurea heteromalla. *Phytomedicine* **2012**, *19*, 988–997. [CrossRef] [PubMed]
2. Sajad, M.A.; Khan, M.S.; Bahadur, S.; Shuaib, M.; Naeem, A.; Zaman, W.; Ali, H. Nickel phytoremediation potential of some plant species of the Lower Dir, Khyber Pakhtunkhwa, Pakistan. *Limnol. Rev.* **2020**, *20*, 13–22. [CrossRef]
3. Khatun, S. Antimicrobial activity of tuber extracts of the medicinal plant coleus forskohlii. *Plant Cell Biotechnol. Mol. Biol.* **2020**, *21*, 11–17.
4. Gao, Q.; Yang, M.; Zuo, Z. Overview of the anti-inflammatory effects, pharmacokinetic properties and clinical efficacies of arctigenin and arctiin from *Arctium lappa* L. *Acta Pharmacol. Sin.* **2018**, *39*, 787–801. [CrossRef]
5. Kang, H.S.; Lee, J.Y.; Kim, C.J. Anti-inflammatory activity of arctigenin from Forsythiae Fructus. *J. Ethnopharmacol.* **2008**, *116*, 305–312. [CrossRef]
6. Hayashi, K.; Narutaki, K.; Nagaoka, Y.; Hayashi, T.; Uesato, S. Therapeutic effect of arctiin and arctigenin in immunocompetent and immunocompromised mice infected with influenza A virus. *Biol. Pharm. Bull.* **2010**, *33*, 1199–1205. [CrossRef] [PubMed]
7. Cheng, X.; Zhong, F.; He, K.; Sun, S.; Chen, H.; Zhou, J. EHHM, a novel phenolic natural product from *Livistona chinensis*, induces autophagy-related apoptosis in hepatocellular carcinoma cells. *Oncol. Lett.* **2016**, *12*, 3739–3748. [CrossRef] [PubMed]
8. Rathore, S.; Tiwari, J.K.; Malik, Z.A. Ethnomedicinal survey of herbaceous flora traditionally used in health care practices by inhabitants of dhundsir gad watershed of garhwal himalaya, india. *Glob. J. Res. Med. Plants Indig. Med.* **2015**, *4*, 65–78.

9. Khare, C.P. (Ed.) *Indian Medicinal Plants*; Springer: New York, NY, USA, 2007; ISBN 978-0-387-70637-5.
10. Kala, C.P. Medicinal plants of the high altitude cold desert in India: Diversity, distribution and traditional uses. *Int. J. Biodivers. Sci. Manag.* **2006**, *2*, 43–56. [CrossRef]
11. Singh, A.; Lal, M.; Samant, S.S. Diversity, indigenous uses and conservation prioritization of medicinal plants in lahaul valley, proposed cold desert biosphere reserve, India. *Int. J. Biodivers. Sci. Manag.* **2009**, *5*, 132–154. [CrossRef]
12. Wahid, M.; Saqib, F.; Ahmedah, H.T.; Gavris, C.M.; De Feo, V.; Hogea, M.; Moga, M.; Chicea, R. *Cucumis sativus* L. Seeds Ameliorate Muscular Spasm-Induced Gastrointestinal and Respiratory Disorders by Simultaneously Inhibiting Calcium Mediated Signaling Pathway. *Pharmaceuticals* **2021**, *14*, 1197. [CrossRef] [PubMed]
13. Sirous, H.; Chemi, G.; Campiani, G.; Brogi, S. An integrated in silico screening strategy for identifying promising disruptors of p53-MDM2 interaction. *Comput. Biol. Chem.* **2019**, *83*, 107105. [CrossRef]
14. Kuhn, B.; Kollman, P.A. Binding of a diverse set of ligands to avidin and streptavidin: An accurate quantitative prediction of their relative affinities by a combination of molecular mechanics and continuum solvent models. *J. Med. Chem.* **2000**, *43*, 3786–3791. [CrossRef]
15. Gilani, A.U.H.; Shah, A.J.; Yaeesh, S. Presence of cholinergic and calcium antagonist constituents in Saussurea lappa explains its use in constipation and spasm. *Phyther. Res.* **2007**, *21*, 541–544. [CrossRef]
16. Chen, J.C.; Ho, T.Y.; Chang, Y.S.; Wu, S.L.; Hsiang, C.Y. Anti-diarrheal effect of Galla Chinensis on the *Escherichia coli* heat-labile enterotoxin and ganglioside interaction. *J. Ethnopharmacol.* **2006**, *103*, 385–391. [CrossRef] [PubMed]
17. Cheng, Y.; Li, X.; Tse, H.F.; Rong, J. Gallic acid-L-leucine conjugate protects mice against LPS-induced inflammation and sepsis via correcting proinflammatory lipid mediator profiles and oxidative stress. *Oxid. Med. Cell. Longev.* **2018**, *2018*, 1081287. [CrossRef]
18. Gilani, A.H.; Ghayur, M.N.; Saify, Z.S.; Ahmed, S.P.; Choudhary, M.I.; Khalid, A. Presence of cholinomimetic and acetylcholinesterase inhibitory constituents in betel nut. *Life Sci.* **2004**, *75*, 2377–2389. [CrossRef] [PubMed]
19. Lanuzza, F.; Occhiuto, F.; Monforte, M.T.; Tripodo, M.M.; D'Angelo, V.; Galati, E.M. Antioxidant phytochemicals of *Opuntia ficus-indica* (L.) Mill. cladodes with potential anti-spasmodic activity. *Pharmacogn. Mag.* **2017**, *13*, S424–S429. [CrossRef]
20. Ghayur, M.N.; Khan, H.; Gilani, A.H. Antispasmodic, bronchodilator and vasodilator activities of (+)-catechin, a naturally occurring flavonoid. *Arch. Pharm. Res.* **2007**, *30*, 970–975. [CrossRef]
21. Zhang, M.; Zhao, C.; Shao, Q.; Yang, Z.; Zhang, X.; Xu, X.; Hassan, M. Determination of water content in corn stover silage using near-infrared spectroscopy. *Int. J. Agric. Biol. Eng.* **2019**, *12*, 143–148. [CrossRef]
22. Saqib, F.; Janbaz, K.H. Rationalizing ethnopharmacological uses of *Alternanthera sessilis*: A folk medicinal plant of Pakistan to manage diarrhea, asthma and hypertension. *J. Ethnopharmacol.* **2016**, *182*, 110–121. [CrossRef] [PubMed]
23. Karaki, H.; Ozaki, H.; Hori, M.; Mitsui-Saito, M.; Amano, K.I.; Harada, K.I.; Miyamoto, S.; Nakazawa, H.; Won, K.J.; Sato, K. Calcium movements, distribution, and functions in smooth muscle. *Pharmacol. Rev.* **1997**, *49*, 157–230. [PubMed]
24. Bolton, T.B. Mechanisms of action of transmitters and other substances on smooth muscle. *Physiol. Rev.* **1979**, *59*, 606–718. [CrossRef]
25. Fleckenstein, A. Specific pharmacology of calcium in myocardium, cardiac pacemakers, and vascular smooth muscle. *Annu. Rev. Pharmacol. Toxicol.* **1977**, *17*, 149–166. [CrossRef]
26. Yakubu, M.T.; Salimon, S.S. Antidiarrhoeal activity of aqueous extract of *Mangifera indica* L. leaves in female albino rats. *J. Ethnopharmacol.* **2015**, *163*, 135–141. [CrossRef] [PubMed]
27. Janbaz, K.H.; Jan, A.; Qadir, M.I.; Gilani, A.H. Spasmolytic, bronchodilator and vasorelaxant activity of methanolic extract of tephrosia purpurea. *Acta Pol. Pharm.-Drug Res.* **2013**, *70*, 261–269.
28. Janbaz, K.H.; Arif, J.; Saqib, F.; Imran, I.; Ashraf, M.; Zia-Ul-Haq, M.; Jaafar, H.Z.E.; De Feo, V. In-vitro and in-vivo validation of ethnopharmacological uses of methanol extract of *Isodon rugosus* Wall. ex Benth. (Lamiaceae). *BMC Complement. Altern. Med.* **2014**, *14*, 71. [CrossRef]
29. Janbaz, K.H.; Nisa, M.; Saqib, F.; Imran, I.; Zia-Ul-Haq, M.; De Feo, V. Bronchodilator, vasodilator and spasmolytic activities of methanolic extract of *Myrtus communis* L. *J. Physiol. Pharmacol.* **2013**, *64*, 479–484.
30. Ghayura, M.N.; Gilani, A.H. A-Adrenergic Receptor Mediated Hypertensive and Vasoconstrictor Effects of Dietary Radish Leaves Extract. *J. Health Sci.* **2007**, *53*, 151–155. [CrossRef]
31. Gilani, A.H.; Jabeen, Q.; Ghayur, M.N.; Janbaz, K.H.; Akhtar, M.S. Studies on the antihypertensive, antispasmodic, bronchodilator and hepatoprotective activities of the *Carum copticum* seed extract. *J. Ethnopharmacol.* **2005**, *98*, 127–135. [CrossRef]
32. Deliorman Orhan, D.; Hartevioğlu, A.; Küpeli, E.; Yesilada, E. In vivo anti-inflammatory and antinociceptive activity of the crude extract and fractions from *Rosa canina* L. fruits. *J. Ethnopharmacol.* **2007**, *112*, 394–400. [CrossRef]
33. Zhao, C.; Cao, Y.; Ma, Z.; Shao, Q. Optimization of liquid ammonia pretreatment conditions for maximizing sugar release from giant reed (*Arundo donax* L.). *Biomass Bioenergy* **2017**, *98*, 61–69. [CrossRef]
34. Qiao, X.; Zhao, C.; Shao, Q.; Hassan, M. Structural Characterization of Corn Stover Lignin after Hydrogen Peroxide Presoaking Prior to Ammonia Fiber Expansion Pretreatment. *Energy Fuels* **2018**, *32*, 6022–6030. [CrossRef]
35. Chen, D.; Cen, K.; Cao, X.; Chen, F.; Zhang, J.; Zhou, J. Insight into a new phenolic-leaching pretreatment on bamboo pyrolysis: Release characteristics of pyrolytic volatiles, upgradation of three phase products, migration of elements, and energy yield. *Renew. Sustain. Energy Rev.* **2021**, *136*, 110444. [CrossRef]
36. Chen, D.; Cen, K.; Cao, X.; Zhang, J.; Chen, F.; Zhou, J. Upgrading of bio-oil via solar pyrolysis of the biomass pretreated with aqueous phase bio-oil washing, solar drying, and solar torrefaction. *Bioresour. Technol.* **2020**, *305*, 123130. [CrossRef] [PubMed]

37. Chen, D.; Chen, F.; Cen, K.; Cao, X.; Zhang, J.; Zhou, J. Upgrading rice husk via oxidative torrefaction: Characterization of solid, liquid, gaseous products and a comparison with non-oxidative torrefaction. *Fuel* **2020**, *275*, 117936. [CrossRef]
38. Chen, D.; Cen, K.; Chen, F.; Ma, Z.; Zhou, J.; Li, M. Are the typical organic components in biomass pyrolyzed bio-oil available for leaching of alkali and alkaline earth metallic species (AAEMs) from biomass? *Fuel* **2020**, *260*, 116347. [CrossRef]
39. VAN ROSSUM, J.M. Cumulative dose-response curves. II. Technique for the making of dose-response curves in isolated organs and the evaluation of drug parameters. *Arch. Int. Pharmacodyn. Thér.* **1963**, *143*, 299–330. [PubMed]
40. Zarei, M.; Mohammadi, S.; Komaki, A. Antinociceptive activity of *Inula britannica* L. and patuletin: In vivo and possible mechanisms studies. *J. Ethnopharmacol.* **2018**, *219*, 351–358. [CrossRef]

plants

MDPI

Article

Chemical Composition and Assessment of Antimicrobial Activity of Lavender Essential Oil and Some By-Products

Alexandru Ciocarlan [1], Lucian Lupascu [1], Aculina Aricu [1], Ion Dragalin [1], Violeta Popescu [1], Elisabeta-Irina Geana [2,3], Roxana Elena Ionete [2], Nicoleta Vornicu [3], Octavian G. Duliu [4,5], Gergana Hristozova [5,6] and Inga Zinicovscaia [1,5,7,*]

1 Department of Chemistry of Natural and Biologically Active Compounds, Institute of Chemistry, Academiei Str. 3, MD-2028 Chisinau, Moldova; algciocarlan@yahoo.com (A.C.); lucian1978@mail.ru (L.L.); aculina.aricu@gmail.com (A.A.); iondragalin@yahoo.com (I.D.); violeta.popescu74@gmail.com (V.P.)
2 Department of Research and Development, National Research and Development Institute for Cryogenics and Isotopic Technologies—ICSI Rm. Valcea, 4th Uzinei Str., PO Raureni Box 7, 240050 Rm. Valcea, Romania; irina.geana@icsi.ro (E.-I.G.); roxana.ionete@icsi.ro (R.E.I.)
3 Metropolitan Center of Research T.A.B.O.R., 9 Closca Str., RO-700066 Iasi, Romania; cmctaboriasi@yahoo.com
4 Department of Structure of Matter, Faculty of Physics, Earth and Atmospheric Physics and Astrophysics, University of Bucharest, 405 Atomistilor Street, 077125 Magurele, Romania; o.duliu@upcmail.ro
5 Department of Nuclear Physics, Joint Institute for Nuclear Research, 6 Joliot-Curie Str., 141980 Dubna, Russia; gerihris2@gmail.com
6 Faculty of Physics and Technology, Plovdiv University "Paisii Hilendarski", 24 Tsar Asen Str., 4000 Plovdiv, Bulgaria
7 Department of Nuclear Physics, Horia Hulubei National Institute for R&D in Physics and Nuclear Engineering, 30 Reactorului Str. MG-6, 077125 Magurele, Romania
* Correspondence: zinikovskaia@mail.ru; Tel.: +7-496-216-5609

Citation: Ciocarlan, A.; Lupascu, L.; Aricu, A.; Dragalin, I.; Popescu, V.; Geana, E.-I.; Ionete, R.E.; Vornicu, N.; Duliu, O.G.; Hristozova, G.; et al. Chemical Composition and Assessment of Antimicrobial Activity of Lavender Essential Oil and Some By-Products. *Plants* **2021**, *10*, 1829. https://doi.org/10.3390/plants10091829

Academic Editors: Marc (Vlaic) Romina Alina, Crina Muresan, Andruta Muresan and Juei-Tang Cheng

Received: 29 July 2021
Accepted: 31 August 2021
Published: 3 September 2021

Abstract: The producers of essential oils from the Republic of Moldova care about the quality of their products and at the same time, try to capitalize on the waste from processing. The purpose of the present study was to analyze the chemical composition of lavender (*Lavanda angustifolia* L.) essential oil and some by-products derived from its production (residual water, residual herbs), as well as to assess their "in vitro" antimicrobial activity. The gas chromatography-mass spectrometry analysis of essential oils produced by seven industrial manufacturers led to the identification of 41 constituents that meant 96.80–99.79% of the total. The main constituents are monoterpenes (84.08–92.55%), followed by sesquiterpenes (3.30–13.45%), and some aliphatic compounds (1.42–3.90%). The high-performance liquid chromatography analysis allowed the quantification of known triterpenes, ursolic, and oleanolic acids, in freshly dried lavender plants and in the residual by-products after hydrodistillation of the essential oil. The lavender essential oil showed good antibacterial activity against *Bacillus subtilis, Pseudomonas fluorescens, Xanthomonas campestris, Erwinia carotovora* at 300 μg/mL concentration, and *Erwinia amylovora, Candida utilis* at 150 μg/mL concentration, respectively. Lavender plant material but also the residual water and ethanolic extracts from the solid waste residue showed high antimicrobial activity against *Aspergillus niger, Alternaria alternata, Penicillium chrysogenum, Bacillus* sp., and *Pseudomonas aeroginosa* strains, at 0.75–6.0 μg/mL, 0.08–0.125 μg/mL, and 0.05–4.0 μg/mL, respectively.

Keywords: *Lavandula angustifolia* L.; essential oil; by-products; terpenic compounds; chromatographic analyses; antimicrobial activity; statistical data analysis

1. Introduction

Lavandula angustifolia Mill. (syn. *Lavandula vera* DC, syn. *Lavandula officinalis* Chaix ex Vill., syn. *Lavandula spica* L.) is a perennial evergreen shrub of the family *Lamiaceae*, native to the Mediterranean region. Nowadays, this species is naturalized almost all over Europe, North Africa, United States, and Australia [1]. *L.angustifolia* (Lavander) is one of the most valuable medicinal and aromatic plants traditionally used to treat pain, parasitic infections, burns, insect bites, cramps, and muscle spasms [2]. In addition to its application in herbal treatment, lavender is also cultivated for the essential oils used in aromatherapy and the cosmetic, food, and flavour industries [3–5].

This is possible due to the presence of a set of biologically active substances, especially in essential oil, which possesses a multidirectional therapeutic activity being used in the treatment of gastrointestinal, cardiovascular, respiratory, and urinary infections [6]. Scientific studies reported anti-inflammatory [7], antioxidant [8,9], sedative [10], cytotoxic [11,12], analgesic [7], antimicrobial [6,13,14], and anticonvulsive [15] properties of *L. angustifolia* essential oil. Literature data reveal a huge variation in terms of *L. angustifolia* essential oil content, with values ranging between 0.5 and 6.25% in the case of essential oil obtained from fresh and dry inflorescences [16]. The main constituents of *L. angustifolia* essential oil are linalool, linalyl acetate, 1,8-cineole, borneol, camphor, lavandulyl acetate, β-caryophyllene, β-ocimene, α-fenchone, terpinen-4-ol, caryophyllene oxide, limonene, pinenes, geranyl acetate, β-farnesene, santalene, lavandulol, camphene, geraniol, and α-terpineol [8,11,13,14,17–26]. The content of oxygenated monoterpenes prevails in *L. angustifolia* essential oil and varies between 36.33 and 92.90% [16].

The therapeutic effects of *L. angustifolia* are also determined by secondary metabolites such as oleanolic and ursolic acids, together with other pentacyclic triterpenes. [27,28]. It has been proven experimentally that both compounds in pure forms, as well as their synthetic derivatives, show multiple biological activities [29–38].

Some by-products, e.g., pomace or solid residues, that resulted after hydrodistillation of essential oil-producing plants could be considered as a source of biologically active compounds such as ursolic and oleanolic acids. In addition, residual distillation waters have various applications due to their aromatic and antimicrobial properties [39–43].

Antibiotic resistance is becoming one of the main problems of modern medicine since it substantially reduces the effectiveness of antibacterial treatments and is linked to increased patient mortality. As a result, known antibacterial preparations cease to be safe and effective against infections caused by resistant bacteria, leading to increasingly serious cases, including hospital-acquired complications. This requires the discovery of new classes of antibiotics or optimization and a combination of known compounds. However, microorganisms will likely evolve resistance in time and further research and development may be hard to sustain by the pharmaceutical companies. For this reason, studies are being conducted to identify effective remedies against multidrug-resistant strains. Preference is given to natural products among which are the essential oils [44], including lavender [45], or their combination with antibiotics [46]. Still, information about the antimicrobial activity of residual water and ethanolic extracts is very scanty and is mainly related to Lavander hydrosol, which is produced synthetically [47].

The aim of this study was to (i) evaluate the chemical composition of lavender essential oil and some of the waste by-products produced industrially in the Republic of Moldova using different chromatographic techniques; (ii) assess the in vitro antimicrobial activity of extracted compounds; and (iii) distinguish, using statistical analysis, between different lavender oils produced in different regions of the Republic of Moldova (Northern, Central, and Southern), based on the terpenic and aliphatic compounds.

2. Results

2.1. GC-MS Analysis Results

A total of 41 constituents of lavender essential oil were identified by means of gas chromatography-mass spectrometry (GC-MS) analysis (Table 1).

It must be mentioned that the essential oil with the richest content was made by producer P1, which is the largest and operates a stationary modern factory. By contrast, producers P2 to P7 use mobile installations and process raw plant material directly in the field, in modernized or artisanal installations, and this may influence the chemical composition of essential oils and resulting by-products.

Table 1. Phytochemical (terpenic and aliphatic compounds) composition of lavender essential oil of Moldovan origin.

No.	RT* (min)	Component	Producer, Content (%)						
			P1	P2	P3	P4	P5	P6	P7
1	4.416	α-Pinene	0.36	0.57	0.36	0.18	0.09	0.26	0.57
2	4.710	Camphene	0.34	0.47	0.30	0.09	0.09	0.26	0.63
3	5.179	Sabinene	0.14	-	-	-	-	-	-
4	5.240	1-Octen-3-ol		0.83	-	-	-	0.15	0.27
5	5.263	β-Pinene	0.51	-	0.34	0.39	0.38	0.22	0.57
6	5.398	Octan-3-one	0.28	0.51	0.31	0.25	0.39	0.12	0.21
7	5.489	β-Myrcene	1.06	1.50	0.96	0.80	0.89	0.62	0.89
8	5.577	Octan-3-ol	0.13	0.18	0.20	0.17	0.20	-	-
9	5.962	*n*-Hexyl acetate	0.59	1.27	0.31	0.42	0.55	0.52	1.17
10	6.284	*p*-Cymene	0.22	0.46	0.22	0.14	0.10	0.24	0.39
11	6.400	Limonene	0.52	1.79	0.79	0.45	0.55	1.17	1.93
12	6.455	1,8-Cineol (eucalyptol)	5.00	3.81	2.22	3.73	4.44	3.83	9.29
13	6.574	(*E*)-Ocimene	8.06	5.87	6.85	7.86	4.37	5.25	7.15
14	6.807	(*Z*)-Ocimene	3.74	3.45	2.59	2.55	1.86	1.76	2.16
15	7.087	γ-Terpinene	0.10	0.29	0.29	0.06	0.07	0.07	0.15
16	7.443	Linalool oxide	-	0.17	0.07	0.11	0.12	-	-
17	7.824	δ-Terpinene	0.27	-	-	-	-	-	-
18	7.825	α-Terpinolene	-	0.58	0.25	0.21	0.32	0.16	0.23
19	8.238	Linalool	23.54	27.98	29.06	25.57	40.68	33.29	26.19
20	8.392	Oct-1-en-3-yl acetate	0.56	0.82	0.60	0.71	0.63	0.39	0.58
21	9.308	Camphor	0.47	0.47	0.37	0.36	0.30	0.32	0.61
22	9.847	Borneol	1.92	2.15	1.68	1.28	1.65	1.41	2.40
23	10.00	(3*E*,5*Z*)-Undeca-1,3,5-triene	0.17	-	-	-	-	-	-
24	10.15	Terpin-1-en-4-ol	1.30	4.65	5.98	0.94	1.67	1.41	1.03
25	10.41	Cryptone	0.29	0.29	-	-	-	0.30	0.33
26	10.50	α-Terpineol	2.42	3.31	2.02	1.42	7.95	1.49	1.61
27	11.49	Nerol	0.38	0.46	0.23	0.14	1.14	-	-
28	11.84	*p*-Cumic aldehyde	0.13	0.15	-	-	-	0.18	-
29	12.32	Linalyl acetate	26.55	20.26	28.65	32.25	16.68	33.30	28.10
30	13.01	Bornyl acetate	0.32	0.25	0.24	0.27	0.19	0.17	0.24
31	13.11	Lavandulyl acetate	4.88	2.84	2.36	4.83	4.78	2.56	3.07
32	14.98	Neryl acetate	0.78	0.91	0.39	0.33	1.53	0.31	0.37
33	15.47	Geranyl acetate	1.31	1.69	0.79	0.73	2.70	0.59	0.67
34	15.66	α-Zingiberene	0.15	-	-	-	-	-	-
35	16.47	β-Caryophyllene	6.25	5.33	4.62	5.44	1.64	4.93	4.32
36	16.80	α-Bergamotene	0.27	0.28	0.19	0.20	0.05	0.16	-
37	17.31	(*E*)-β-Farnesene	4.86	2.59	3.65	3.94	1.23	2.46	2.45
38	17.97	β-Cubebene	1.12	0.82	1.03	0.69	0.17	-	-
39	18.75	γ-Cadinene	0.18	0.53	-	-	-	0.68	0.65
40	20.39	Caryophyllene oxide	0.45	0.69	0.19	0.29	0.21	0.35	0.28
41	21.69	Cadinol	0.17	0.57	-	-	-	0.18	-
	Total content, (%)		99.80	98.79	98.17	96.80	97.62	99.11	98.51

*RT: Retention time; P 1–7: Producers.

According to the GC-MS data, the chemical composition of lavender essential oil produced in Moldova consisted mainly of terpenic and aliphatic compounds and their content varied within the limits indicated in Table 2.

The GC-MS analysis of extracts from residual waters (RW) showed that they contained only several hydrophilic components (see Section 3.3) and represented about 0.3–0.5% of the volume.

2.2. RP-HPLC Analysis Results

The content of triterpenic oleanolic acid (OA) and ursolic acid (UA) was established in freshly dried lavender plants and in dried solid residues (after hydrodistillation) via RP-HPLC analysis.

Table 2. Chemical composition of lavender essential oil.

Class	Subclass	Content, (%)
Terpenic compounds		94.89–97.77
	Monoterpenes	84.08–92.55
	Monoterpene hydrocarbons	8.72–15.32
	Oxygenated monoterpenes	69.00–83.83
	Sesquiterpenes	3.30–13.45
	Sesquiterpene hydrocarbons	3.09–12.83
	Oxygenated sesquiterpenes	0.19–1.26
Aliphatic compounds		1.42–3.90
	Hydrocarbons	0.17
	Alcohols	0.13–1.01
	Ketones	0.25–0.80
	Esters	0.91–2.09
	Total	96.80–99.79

The results were expressed as mg/g for extracts and mg/100 g for the ratio plant material/solid residue (Tables 3 and 4). It was observed that fresh plants had a much higher content of OA and UA.

Table 3. The OA and UA content of lavender plant material (DW).

Lavender Plant Material	Extract Yield (%)	Concentration (mg/g Extract)		Concentration (mg/100 g Lavender Plant Material, DW)	
		OA	UA	OA	UA
LPM 1	9.94	16.19	37.46	160.95	372.36
LPM 2	8.83	19.09	60.82	168.57	537.00
LPM 3	9.91	13.43	33.28	133.11	329.83

Table 4. The OA and UA content of lavender by-product (solid waste residue), (DW).

Lavender by-Product (Solid Residue, SR)	Extract Yield (%)	Concentration (mg/g Extract)		Concentration (mg/100 g Dry Solid Residue)	
		OA	UA	OA	UA
SR 1	3.88	29.21	80.82	113.47	313.95
SR 2	3.68	39.37	135.56	144.98	499.15
SR 3	4.15	27.48	87.90	114.07	364.89

The lower content of OA and UA in solid residues can be explained by their loss and derivatization/degradation during hydrodistillation in an aqueous medium at elevated temperatures (Table 4). The latter seemed more relevant since neither OA nor UA was found in residual water extracts (see Section 3.3).

2.3. Microbial Inhibition Assessment Results

The microbial activity assessment of lavender essential oil extracts from lavender plant material (LPM), lavender by-products (residual water (RW), and solid waste residue (SR)) was performed by serial dilution methods against several non-pathogenic Gram-positive and Gram-negative bacteria strains and fungi species, including phytopathogenic ones (e.g., *Xanthomonas campestris*, *Erwinia amylovora*, and *Erwinia carotovora*).

The results of the lavender essential oil antibacterial and antifungal activity tests are presented in Table 5.

Table 5. The antimicrobial activity of lavender essential oil.

Sample	MBC and MFC, μg/mL					
	Bacillus subtilis	*Pseudomonas fluorescens*	*Xantdomonas campestris*	*Erwinia amylovora*	*Erwinia carotovora*	*Candida utilis*
LEO	300	300	300	150	300	150

MBC: Minimal bactericidal concentration; MFC: Minimal fungicidal concentration.

The same method was applied for residual waters, ethanolic extracts from solid residues, and freshly dried lavender plant materials (Table 6).

Table 6. The antimicrobial activity of residual water and ethanolic extracts from lavender plants.

Sample	MIC (μg/mL)				
	Aspergillus niger	*Alternaria alternata*	*Penicillium chrysogenum*	*Bacillus* sp.	*Pseudomonas aeruginosa*
Residual Water	0.08	0.08	0.08	0.125	0.125
Extract from SR	0.50	0.50	0.50	4	4
Extract from LPM	0.75	0.75	0.75	6	6
Caspofungin [a]	0.24	0.24	0.24	-	-
Kanamycin [b]	-	-	-	3.5	3.5

RSD (μg/mL): [a] ±0.001 [b] ±0.0002.

All of the samples were preliminarily tested for their in vitro antimicrobial activity and antifungal effect against pure cultures of three species of fungi (*Aspergillus niger, Alternaria alternate, Penicillium chrysogenum*) and against Gram-positive (*Bacillus* sp.) and Gram-negative bacteria (*Pseudomonas aeruginosa*). Microorganisms were provided by the American Type Culture Collection (ATCC, USA). Caspofungin and Kanamycin were used as performance standards for testing the antifungal and antibacterial activities. The minimum inhibitory concentration values (MIC) for all the samples and standards are summarized in Table 6.

3. Discussion

3.1. Chemical Composition of Lavender Essential Oils

The essential oil manufactured by producer P1, destined for export, had the following physico-chemical properties: Density (20 °C)—0.8920 g/mL; refractive index ($n^{20}{}_D$)—1.4660, and optical rotation ($\alpha^{20}{}_D$)— −7.0°.

The most multitudinous group of terpenic compounds are monoterpenes, which include C_{10}-hidrocarbones (8.72–15.32%) and their oxygenated derivatives (69.0–83.83%). The main constituents of this group which determine the quality and genuineness of lavender essential oil, according to the International Standard [48], are (%): 1,8-cineol (eucalyptol) (<1.0), (*E*)-ocimene (4.0–10.0), (*Z*)-ocimene (1.5–6.0), linalool (25.0–38.0), camphor (<0.5), terpin-1-en-4-ol (2.0–6.0), α-terpineol (<1.0), linalyl acetate (25.0–45.0), and lavandulyl acetate (>2.0) (Tables 1 and 2).

The content of sesquiterpene hydrocarbons and their oxygenated derivatives is reported to be within the limits of 3.09–12.83% and 0.19–1.26%, respectively. According to the same source [48], the most important sesquiterpenes are: β-caryophyllene (4.78%), (*E*)-β-farnesene (1.52%), and caryophyllene oxide (0.36%) (Tables 1 and 2).

Aliphatic compounds are of lesser concentration (1.42–3.90%) and in [48] are mentioned: 1-octen-3-ol (0.33%) and octan-3-one (<2.0%) (Tables 1 and 2).

3.2. Chemical Composition of Lavender Plant Material

For the selective extraction of ursolic and oleanolic triterpene acids from the lavender plant materials (LPM), the extraction yield varied between 8.83–9.94%, with the OA content between 13.43–19.09 mg/g and UA content between 33.28–60.82 mg/g. The content of OA and UA in dry (DW) LPM was in the range of 133.11–168.57 mg/100 g, and respectively 329.83–537.00 mg/100 g DW LPM (Table 3).

Moreover, the experimental results showed that the sum of isomeric OA and UA in LPM was about 5% of the DW, in a 1:3.7 ratio, confirming that lavender is a valuable source of natural OA and UA triterpene acids.

3.3. Chemical Composition of Lavender by-Products

The GC-MS analysis of etheric extracts of residual water (RW) proved that they contain hydrophilic monoterpenic compounds such as 1,8-cineol (eucalyptol, 6.31%), linalool oxide (3.08%), linalool (78.05%), terpin-1-en-4-ol (1.92%), and α-terpineol (10.64%).

HPLC quantification of UA and OA indicated that RWs did not contain OA and UA triterpene acids.

In the case of solid waste residues (SR), the average extraction yield was about 3.91%, with the OA content between 27.48–39.37 mg/g and UA content between 80.82–135.56 mg/g (Table 4). The isomeric OA and UA in DW SR ranged between 113.47–144.98 and 313.95–499.15 mg/100 g, respectively (Table 4), with their amount accounting to about 1% of DW, in a 1:3.1 ratio, indicating that lavender by-products are a promising source of OA and UA triterpene acids.

Our results are consistent with other literature data reporting DW of lavender SR values between 136.0–259.7 and 346.3–648.4 mg/100 g [49].

3.4. Antimicrobial Assessments

Phytopathogenic bacteria can cause various diseases of agricultural plants, especially the genera Erwinia and Xanthomonas. For example, *Erwinia amylovora*, the Gram-negative bacterium of the Enterobacteriaceae family, is the causative agent of fire blight, a devastating plant disease that affects a wide range of species of the family Rosaceae and is a major global threat to commercial apple and pear production. [50]. Another species, *E. carotovora*, causes bacterial soft rot in economically important crops, such as potatoes, tomatoes, and cucumbers. In the case of potatoes, the soft rot of the stem and tubers occurs even after harvest, thus considerably reducing the yield [51]. *Xanthomonas campestris* pv. *vesicatoria* is a biotrophic Gram-negative bacterium and is the agent that causes bacterial leaf scorch on tomatoes (*Solanum lycopersicum* L.) and peppers (*Capsicum annuum*), a disease that is present worldwide. Symptoms of bacterial infection include defoliation and chlorotic necrotic lesions on leaves, stems, fruits, and flowers, which subsequently lead to reduced fruit yield [52].

The species *Bacillus subtilis* and *Pseudomonas fluorescens* do not cause any disease to plants but were selected as reference bacteria from the Gram-positive and Gram-negative groups. They are also very suitable as test objects for evaluating the antibacterial activity of the lavender extract. *Candida utilis* and *Saccharomyces cerevisiae* are also non-pathogenic but were used as representatives of the yeast-fungus group for evaluating the antifungal activity of the extract.

It should be mentioned that there is a lack of information about any antimicrobial effects of lavender essential oil on *E. carotovora*, *E. amylovora*, and *C. utilis*.

The in vitro assessment of lavender essential oil of Moldovan origin showed good antibacterial activity against both non-pathogenic Gram-positive/Gram-negative bacteria (*B. subtilis* and *P. fluorescens*) at MBC of 300 µg/mL and good to high antifungal activity against phytopathogenic bacteria (*X. campestris*, *E. amylovora*, *E. carotovora*) and *C. utilis* fungi at MFC of 150–300 µg/mL (Table 5).

The highest antifungal and antibacterial activities were observed for residual water (RW) at 0.08 and 0.125 µg/mL, respectively. Good antifungal and antibacterial activities were ascertained for the SR extract as well (0.50 and 4 µg/mL). The LPM extract showed moderate antifungal and antibacterial activity (0.75 and 6 µg/mL).

The two techniques employed for testing both the disc diffusion and the dilution methods have been developed to yield accurate measurements of antibacterial and antifungal activities and are routinely used in antimicrobial susceptibility testing.

According to the obtained results, the antibacterial activity was similar but the antifungal activity was slightly different, thus suggesting that the activity against different microorganisms could be caused by different components of the oil.

3.5. Statistical Data Analysis

Univariate as well as multivariate statistical data analysis (SDA) represent one of the most reliable methods that permit extracting useful information and inferring different hypotheses concerning the considered set of data. Given the great diversity of organic compounds which can be found in lavender essential oil, multivariate statistical data analysis was an appropriate method allowing to group samples, in this case, according to the lavender oil producer and based on the concentrations of organic compounds (R mode), or, to classify an experimentally determined organic compound based on the concentration in samples (Q mode) [53,54].

It is worth mentioning that, to avoid any errors induced by missing data, SDA was applied only in the cases of compounds with a non-negligible variation present in all the samples (Table 1), i.e., the compounds which permitted generating the box plots in Figure 1a,b.

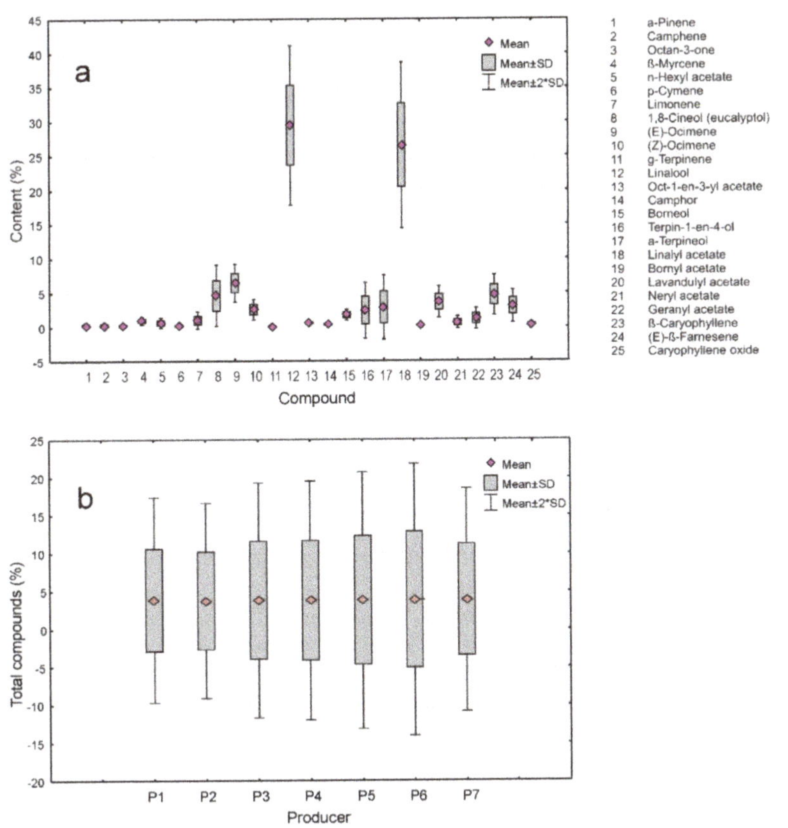

Figure 1. Box plots representing the distribution of (**a**) 25 components of lavender oil and (**b**) total content of compounds present in all the samples (producers).

Univariate SDA was useful in establishing the extent to which the samples of lavender oil by the seven producers were similar. This information was obtained by analyzing the box plot shown in Figure 1a. It was observed that all the samples were quite similar.

To confirm this, we used more univariate tests, such as one-way ANOVA, Tukey's pairwise test, Kruskal-Wallis test of equal medians, as well as Mann-Whitney U tests. All of them confirmed that between the lavender oil samples there are no statistically significant differences. For this reason, we have proceeded with multivariate SDA.

Within multivariate SDA, each sample (case) is characterized by independent parameters (variables), so that the final analysis can be performed in R mode (to study relations between samples based on variables) or Q mode (to study the interrelations between variables based on samples). As both methods were based on the same set of samples and variables, R and Q modes could be considered complementary, which significantly enhanced the analysis.

Depending on the situation, cases/variables can be grouped by a multitude of procedures among which covariance and correlation are frequently utilized.

In the case of lavender samples, the best results were obtained by the principal component analysis (PCA) applied in both R and Q modes. With respect to the other two SDA methods, cluster analysis and K mean clustering, PCA permitted evidencing the association of samples, i.e., seven producers of lavender oil in R mode, as well as 25 lavender oil compounds in Q mode. Moreover, in R mode, a tree diagram corresponding to the cluster analysis (Euclidean distances) is, concerning the number and structure of clusters, similar to PCA based on correlation. For this reason, we restrained our SDA to both R and Q mode PCA.

The results, represented by the principal component (PC) 2 vs. PC 1 bi-plots, are illustrated in Figure 2a,b, respectively. In both cases, the PCA was based on correlations between variables (organic compounds, R mode) or samples (lavender oil producer, Q mode). Moreover, the loadings of each variable or sample were represented by Factor 2 vs. Factor 1 bi-plots in the corresponding insets: Variables in Figure 2a and samples in Figure 2b.

Accordingly, the result of PCA in R mode is illustrated by the bi-plot in Figure 1a. The existence of at least three clusters can be remarked, two of which consist of only one member, i.e., producers P2 and P5, and a third one, grouping the rest of the producers. The bi-plot illustrating the contribution of each compound to the PC1 and PC2 showed a relatively balanced situation, as both Factors 1 and 2 had similar contributions to PC, consisting of 36.46 and 27.25%, respectively. It is worth mentioning that a similar result was obtained by considering the PC3 vs. PC2, which most probably could be explained by their contribution to the total variance, 25.25 and 17.57%, respectively. The corresponding screen plot in Figure 3a illustrated this finding.

Complementary to the R-mode, a Q mode PC2 vs. PC1 bi-plot, shown in Figure 2b, consisted of three clusters, two of which contained a single organic compound, i.e., linalyl acetate and linalool, while the third one included all other 23 compounds. This result was in good agreement with the composition of the investigated samples, according to which, both linalyl acetate and linalool were characterized by the highest concentrations and variances.

On the contrary, Factor 2 vs. Factor 1 (Figure 2b, inset), except for Producers 2 (P2) and 5 (P5), were nearly coincident and negatively oriented along the first axis, which suggested an almost equivalent contribution to the total variance. This finding may explain the fact that PC1 contributed about 96% to the total variance, as shown in the corresponding screen plot (Figure 3b). In this regard, it is of interest to remark, as mentioned before, that P2 and P5 formed two different uni-component clusters (Figure 2b).

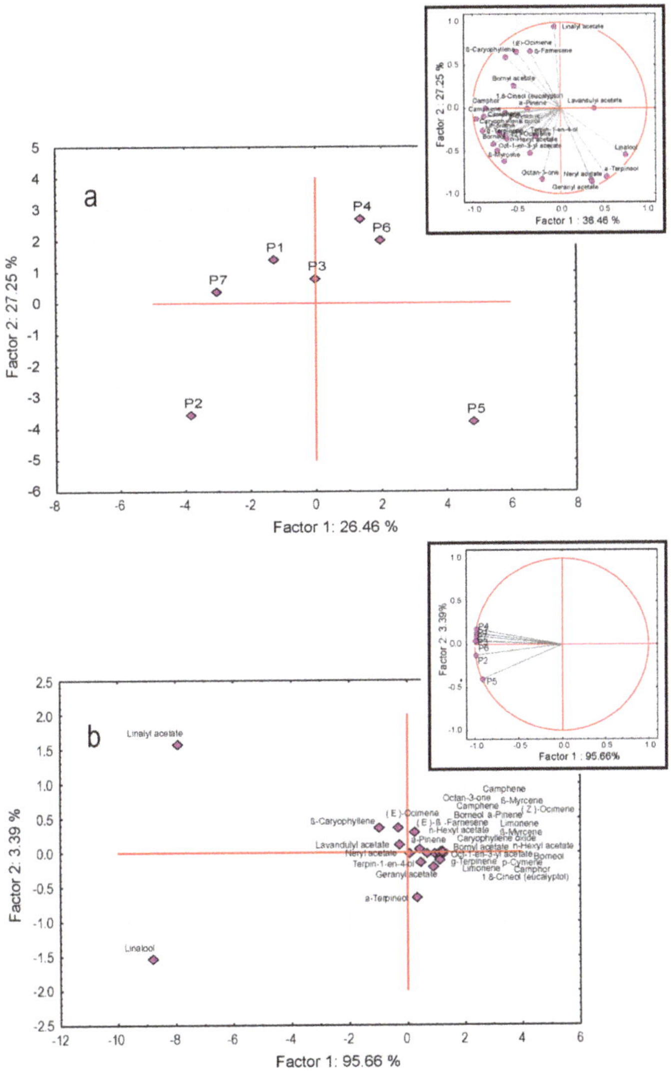

Figure 2. The results of R (**a**) and Q (**b**) mode PCA. The insets illustrate the contribution of the corresponding principal component (PC) analysis.

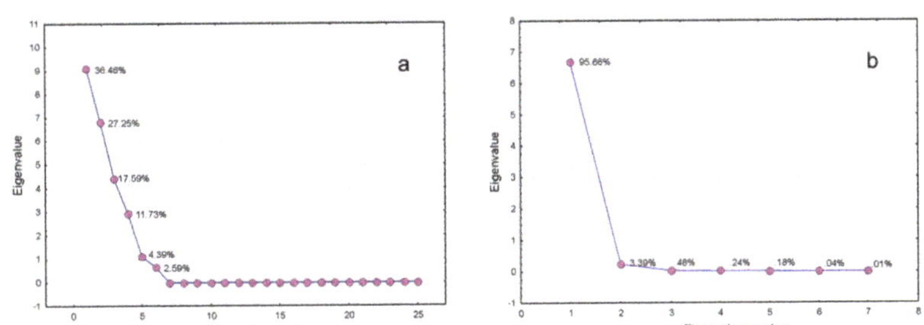

Figure 3. The screen plots corresponding to R-mode (**a**) and Q-mode (**b**) PCA.

4. Materials and Methods

4.1. Samples Collection

The samples of *L. angustifolia* vegetal raw material, by-products, as well as the main product—lavender essential oil (LEO), were provided between 2016 and 2018 by seven producers (P 1-7) from different regions of the Republic of Moldova (Northern, Central, and Southern): P1—Causeni district; P2—Donduseni district; P3 and P6—Rezina district; P4—Falesti district; P5—Dubasari district; and P7—Ungheni district.

For OA and UA characterization, fresh lavender inflorescences were collected directly from the lavender fields near the Pervomaisc village, Causeni district (46°42′04″ N 29°05′21″ E). The inflorescences were dried in shaded places to obtain lavender plant material samples (LPM) (n = 3) which were subjected to HPLC characterization. The by-products which resulted after hydrodistillation (solid residue—SR (n = 3) and residual water—RW (n = 1)), were collected from the factories, dried, and bottled.

4.2. Chemicals

All of the used solvents, reagents, and standards were of analytical grade. Anhydrous sodium carbonate, aluminium chloride, sodium acetate, 96% ethanol, methanol, diethyl ether, and petroleum ether were obtained from Merck (Darmstadt, Germany). Deionized water produced by a Milli-Q Millipore system (Bedford, MA, USA) was used for the preparation of aqueous solutions and UHPLC mobile phases.

The standards used for HPLC-PDA analysis (ursolic and oleanolic acids) were HPLC purity and purchased from Sigma-Aldrich (Steinheim, Germany). Stock solutions of all the standards were prepared in methanol. Working standards were made by diluting the stock solutions in the same solvent. Both stock and working standards were stored at 4 °C until further use.

4.3. Extracts Preparation

The selective extraction of triterpenic ursolic and oleanolic acids with ethanol from the LPM and SR was performed using a Soxhlet type extractor after degreasing with light petroleum ether (b.p. 40 °C) (Figure 4.). The ethanolic extracts were evaporated to dryness at 35 °C under reduced pressure using a rotary evaporator. For HPLC analysis, aliquots of each crude extract were dissolved in methanol using ultrasonication and filtered through a 0.45 μm micro-filter. The extraction and HPLC analysis were performed in duplicate for each plant material and the results were expressed as a mean value. For GC-MS analysis, the industrially produced essential oil samples were dissolved in hexane. The RWs were extracted with diethyl ether and the obtained extracts were subjected to GC chromatographic analysis.

4.4. Analytical GC-MS Analysis

The GC-MS analysis was performed using an Agilent Technologies 7890A gas chromatograph coupled with a 5975C Mass-Selective Detector (MSD) equipped with a split/splitless injector (1 μL). The analysis was carried out on an HP-5MS fused silica capillary calibrated column (30 m × 0.25 mm i.d.; film thickness 0.25 μm). The injector and detector temperatures were kept at 250 °C. Helium was used as carrier gas at a flow rate of 1.1 mL/min; oven temperature program was 70 °C/2 min, which was then programmed to 200 °C at the rate of 5 °C/min, and finally to 300 °C at the rate of 20 °C/min. The split ratio was 1:50, the MSD ionization energy was 70 eV, scan time was 1 s, the acquisition mass was in the range from 30 to 450 amu, and the solvent delay was 3 min.

4.5. Analytical RP-HPLC Analysis

The ursolic and oleanolic acids were quantified by an HPLC-PDA method previously reported [55], using a Thermo Finnigan Surveyor Plus HPLC System (Thermo Fisher Scientific Inc., San Jose, CA, USA). The OA and UA from extracts were identified by their retention time and spectral data by comparison with standards. To confirm the peak identity among possible interference peaks, the technique of standard addition to the sample was applied. Moreover, the peak purity for the interest peaks was satisfactory.

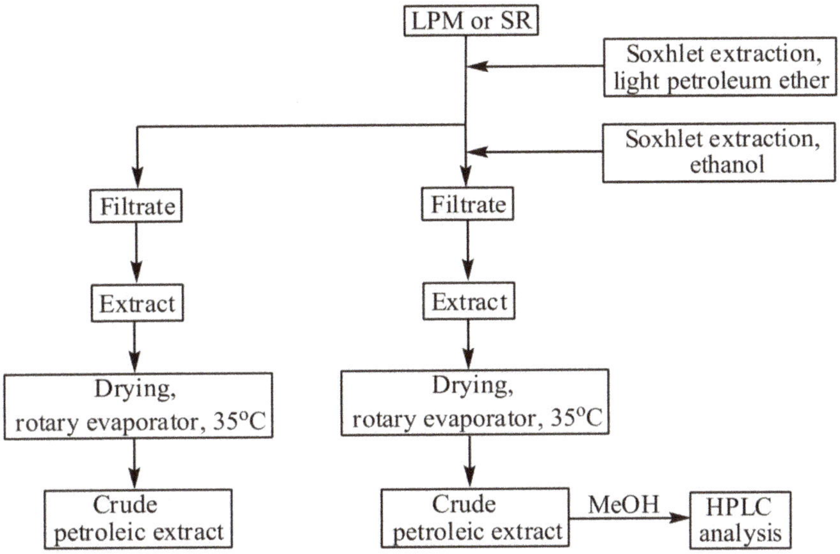

Figure 4. Flowchart of the UA and OA extract preparation.

Calibration curves of the standards covered the range of 1–400 mg/L for both OA and UA and revealed good linearity, with correlation coefficients higher than 0.995 (0.9989 for OA and 0.9991 for UA) [56]. The accuracy of the method (%) was evaluated for spiked samples at 50 mg/L concentration and the obtained average values were 4.31% for OA and 3.65% for UA.

4.6. Antimicrobial Activity Assessment

The in vitro antimicrobial activity tests of methanolic extracts from the SR and RW against three species of fungi (*Aspergillus niger*, *Alternaria alternata*, and *Penicillium chrysogenum*, ATCC 53346, 8741, and 20044) and two species of bacteria (*Pseudomonas aeroginosa* and *Bacillus* sp., ATCC 27813 and 15970) were performed using a previously reported method [57].

Antimicrobial activity assessment of the industrially obtained lavender essential oil samples was performed in vitro on the following microorganisms: Non-pathogenic Gram-positive and Gram-negative strains of *Bacillus subtilis* NCNM BB-01 (ATCC 33608) and *Pseudomonas fluorescens* NCNM-PFB-01 (ATCC 25323), phytopathogenic strains of *Xanthomonas campestris* NCNM BX-01 (ATCC 53196), *Erwinia amylovora* NCNM BE-01 (ATCC 29780), *E. carotovora* NCNM BE-03 (ATCC 15713), and fungus strains of *Candida utilis* NCNM Y-22 (ATCC 44638) and *Saccharomyces cerevisiae* NCNM Y-20 (ATCC 4117) following a method described elsewhere [58].

The compounds Caspofungin and Kanamycin, both from Liofilchem (Roseto degli Abruzzi, Italy), were used as standards for antifungal and antibacterial activity tests.

4.7. Statistical Analysis

All statistical data analyses were performed using the StatSoft Statistica 10 software.

5. Conclusions

More than 40 main constituents of lavender essential oil from seven Moldavian producers were quantified by means of chromatographic and statistical analyses. The experimental data for lavender plant material and solid waste residue proved the possibility of their use as sources of biologically active compounds, such as OA and UA triterpene acids. All of the subjects in the present study, essential oil, residual distillation waste water, and extracts from the solid waste residues have shown high antimicrobial activity against 11 strains of bacteria and fungi, including phytopathogenic ones.

Author Contributions: Conceptualization, A.A. and I.Z.; microbiological assessments, L.L. and N.V.; GC-MC analysis, I.D.; sample preparation, V.P.; HPLC analysis, E.-I.G.; statistical analysis, O.G.D.; data curation, O.G.D.; writing—original draft preparation, A.C.; writing—review and editing, R.E.I., G.H. and I.Z. All authors have read and agreed to the published version of the manuscript.

Funding: This research was funded by the National Agency for Research and Development (ANCD), project PLANTERAS 20.80009.8007.03.

Acknowledgments: O.G.D. and I.Z. wish to acknowledge that their contribution was done within cooperation protocols no. 4920-4-20/22 between the University of Bucharest and the Joint Institute for Nuclear Research, Dubna, Russian Federation, represented by the Frank Laboratory of Neutron Physics. L.L. is grateful to the National Collection of Non-Pathogenic Microorganisms at the Institute of Microbiology and Biotechnology as well as the Laboratory of the Phytopathology and Biotechnology at the Institute of Genetics, Physiology, and Plant Protection for kindly providing the bacterial strains.

Conflicts of Interest: The authors declare no conflict of interest.

Abbreviations

ATCC	American type culture collection
DW	Dry weight
DW PLM	Dry weight lavender plant material
GC-MS	Gas chromatography-mass spectrometry
HPLC	High-performance liquid chromatography
HPLC-PDA	High-performance chromatography-photodiode array detection
LEO	Lavender essential oil
LPM	Lavender plant material
MBC	Minimum bactericidal concentration
MFC	Minimum fungicidal concentration
MIC	Minimum inhibitory concentration
NCNM	National Collection of non-pathogenic microorganisms
OA	Oleanolic acid
RP-HPLC	Reversed-phase high-performance chromatography
RSD	Relative standard deviation
RW	Residual water
SDA	Statistical data analysis
SR	Solid residue
UA	Ursolic acid

References

1. Basch, E.; Foppa, I.; Liebowitz, R.; Nelson, J.; Smith, M.; Sollars, D.; Ulbricht, C. Monograph from National Standard: Lavender (*Lavandula angustifolia* Miller). *J. Herb. Pharmacother.* **2004**, *4*, 63–78. [PubMed]
2. Denner, S.S. Lavandula angustifolia Miller. *Holist. Nurs. Pract.* **2009**, *23*, 57–64. [CrossRef]
3. Da Porto, C.; Decorti, D.; Kikic, I. Flavour compounds of *Lavandula angustifolia* Mill. To use in food manufacturing. Comparison of three different extraction methods. *Food Chem.* **2009**, *112*, 1072–1078. [CrossRef]
4. Erland, L.A.E.; Mahmoud, S.S. Lavender (*Lavandula angustifolia*) Oils. In *Essential Oils in Food Preservation, Flavor and Safety*; Preedy, V.E., Ed.; Academic Press: Amsterdam, The Netherlands; pp. 501–507.
5. Smigielski, K.; Sikora, M.; Majewska, M.; Raj, A. The application of essentials oils to natural and organic cosmetics. *Pol. J. Cosmet.* **2008**, *11*, 89–107.
6. Prusinowska, R.; Smigielski, K. Composition, biological activity and therapeutic effects of lavander (*Lavandula angustifolia* L.): A review. *Herba Pol.* **2014**, *60*, 56–66. [CrossRef]

7. Hajhashemi, V.; Ghannadi, A.; Sharif, B. Anti-inflammatory and analgesic properties of the leaf extracts and essential oil of *Lavandula angustifolia* Mill. *J. Ethnopharm.* **2003**, *89*, 67–81. [CrossRef]
8. Hamad, K.J.; Al-Shaheen, S.J.A.; Kaskoos, R.A.; Ahanad, J.; Jameel, M.; Mir, S.R. Essential oil composition and antioxidant activity of *Lavandula angustifolia* from Irak. *Int. Res. J. Pharm.* **2013**, *4*, 117–120.
9. Djenane, D.; Aıder, M.; Yanguela, J.; Idir, L.; Gomez, D.; Roncales, P. Antioxidant and antibacterial effects of *Lavandula* and *Mentha* essential oils in minced beef inoculated with *E. coli O157:H7* and *S. aureus* during storage at abuse refrigeration temperature. *Meat Sci.* **2012**, *92*, 667–674. [CrossRef] [PubMed]
10. Adaszynska-Skwirzynska, M.; Swarcewicz, M.; Dobrowolska, A. The potential of use lavender from vegetable waste as effective antibacterial and sedative agents. *Med. Chem.* **2014**, *4*, 734–737.
11. Mantovani, A.L.L.; Vieira, G.P.G.; Cunha, W.R.; Groppo, M.; Santos, R.A.; Rodrigues, V.; Magalhaes, L.G.; Crotti, A.E.M. Chemical composition, antischistosomal and cytotoxic effects of the essential oil of *Lavandula angustifolia* grown in South-eastern Brazil. *Rev. Braz. Farmacogn.* **2013**, *23*, 877–884. [CrossRef]
12. Shou-Dong, S.; Chang-Xu, C.; Ji-Shu, Q.; Ming-Hua, S. Study on antitumor effect of *Lavender angustifolia* extract. *Food Sci. Technol.* **2009**, *2*, 213–215.
13. Martucci, J.F.; Gende, L.B.; Neira, L.M.; Ruseckaite, R.A. Oregano and lavender essential oils as antioxidant and antimicrobial additives of biogenic gelatine films. *Ind. Crop. Prod.* **2015**, *71*, 205–213. [CrossRef]
14. Tarek, N.; Hassan, H.M.; AbdelGhani, S.M.M.; Radwan, I.A.; Hammouda, O.; El-Gendy, A.O. Comparative chemical and antimicrobial study of nine essential oils obtained from medicinal plants growing in Egypt. *Beni-Seuf Univ. J. Appl. Sci.* **2014**, *3*, 149–156. [CrossRef]
15. Yamada, K.; Mimaki, Y.; Sashida, Y. Anticonvulsive effects of inhaling lavender oil vapour. *Biol. Pharm. Bull.* **1994**, *17*, 359–360. [CrossRef]
16. Aprotosoaie, A.C.; Gille, E.; Trifan, A.; Luca, V.S.; Miron, A. Essential oil of *Lavandula* genus: A systematic review of their chemistry. *Phytochem. Rev.* **2017**, *16*, 761–799. [CrossRef]
17. Smigielski, K.; Raj, A.; Krosowiak, K.; Grusca, R. Chemical composition of the essential oil of *Lavandula angustifolia* cultivated in Poland. *J. Essent. Oil-Bear. Plants* **2009**, *12*, 338–347. [CrossRef]
18. Hassanpouraghdam, M.B.; Hassani, A.; Vojodi, L.; Hajisamadi, A.B.; Rostami, A. Essential oil constituents of *Lavandula angustifolia* Chaix from Northwest Iran. *Chemija* **2011**, *22*, 167–171.
19. Belhadj, M.M.; Kabouche, A.; Abaza, I.; Aburjai, T.; Tauzani, R.; Kabouche, Z. Chemotypes investigation of *Lavandula* essential oils growing at different North African soils. *J. Mater. Environ. Sci.* **2014**, *5*, 1896–1901.
20. Cong, Y.; Abulizi, P.; Zhi, L.; Wang, X. Chemical composition of the essential oil of *Lavandula angustifolia* from Xinjiang, China. *Chem. Nat. Compd.* **2008**, *44*, 810. [CrossRef]
21. Renaud, E.N.C.; Denys, C.J.; Simon, J.E. Essential oil quantity and composition from 10 cultivars of organically grown lavender and lavandin. *J. Essent. Oil Res.* **2001**, *13*, 269–273. [CrossRef]
22. Jianu, C.; Pop, G.; Gruia, A.T.; Horhat, F.G. Chemical composition and antimicrobial activity of essential oils of lavender (*Lavandula angustifolia*) and lavandin (*Lavandula x intermedia*) grown in western Romania. *Int. J. Agric. Biol.* **2013**, *15*, 772–776.
23. Zagorcheva, T.; Stanev, S.; Rusanov, K.; Atanassov, I. Comparative GC/MS analysis of lavender (*Lavandula angustifolia* Mill.) inflorescence and essential oil volatiles. *Agric. Sci. Technol.* **2013**, *5*, 459–462.
24. Chatzopolou, P.S.; Goliaris, A.H.; Katsiotis, T. Contribution to the analysis of the volatile constituents from some lavender and lavandin cultivars grown in Greece. *Sci. Pharm.* **2003**, *71*, 229–234.
25. Raina, A.P.; Negi, K.S. Comparative essential oil composition of *Lavandula* species from India. *J. Herb Spice Med. Plants* **2012**, *18*, 268–273. [CrossRef]
26. Verma, R.S.; Rahman, L.U.; Chanotiya, C.S.; Verma, R.K.; Chauhan, A.; Singh, A.; Yadav, A.K. Essential oil composition of *Lavandula angustifolia* Mill. cultivated in the mid hills of Uttarakhand. India. *J. Serbian Chem. Soc.* **2010**, *75*, 343–348. [CrossRef]
27. Jaeger, S.; Trojan, H.; Kopp, T.; Laszczyk, M.N.; Scheffler, A. Pentacyclic triterpene distribution in various plants-rich sources for a new group of multi-potent plant extracts. *Molecules* **2009**, *14*, 2016–2031. [CrossRef] [PubMed]
28. Janicsak, G.; Veres, K.; Zoltan Kakasy, A.; Mathe, I. Study of the oleanolic and ursolic acid contents of some species of the Lamiaceae. *Biochem. Syst. Ecol.* **2006**, *34*, 392–396. [CrossRef]
29. Duke, J.A. *Handbook of Biologically Active Phytochemicals and Their Activities*, 1st ed.; CRC Press: Boca Raton, FL, USA, 1992; p. 208.
30. Liu, J. Pharmacology of oleanolic acid and ursolic acid. *J. Ethnopharmacol.* **1995**, *49*, 57–68. [CrossRef]
31. Wolska, K.; Grudniak, A.; Fiecek, B.; Kraczkiewicz-Dowjat, A.; Kurek, A. Antibacterial activity of oleanolic and ursolic acids and their derivatives. *Open Life Sci.* **2010**, *5*, 543–553. [CrossRef]
32. Hsu, Y.L.; Kuo, P.L.; Lin, C.C. Proliferative inhibition, cell-cycle dysregulation, and induction of apoptosis by ursolic acid in human non-small cell lung cancer A549 cells. *Life Sci.* **2004**, *75*, 2303–2316. [CrossRef]
33. Chiang, L.C.; Chiang, W.; Chang, M.Y.; Ng, L.T.; Lin, C.C. Antileukemic activity of selected natural products in Taiwan. *Am. J. Chin. Med.* **2003**, *31*, 37–46. [CrossRef]
34. Zhang, W.; Hong, D.; Zhou, Y.; Zhang, Y.; Shen, Q.; Li, J.; Hu, L.; Li, J. Ursolic acid and its derivative inhibit protein tyrosine phosphatase 1B, enhancing insulin receptor phosphorylation and stimulating glucose uptake. *Biochim. Biophys. Acta* **2006**, *1760*, 1505–1512. [CrossRef] [PubMed]
35. Tian, Z.; Lin, G.; Zheng, R.X.; Huang, F.; Yang, M.S.; Xiao, P.G. Anti-hepatoma activity and mechanism of ursolic acid and its derivatives isolated from Aralia decaisneana. *World J. Gastroenterol.* **2006**, *12*, 874–879. [CrossRef]

36. Kim, Y.K.; Yoon, S.K.; Ryu, S.Y. Cytotoxic Triterpenes from Stem Bark of Physocarpus intermedius. *Planta Med.* **2000**, *66*, 485–486. [CrossRef] [PubMed]
37. Kashiwada, Y.; Nagao, T.; Hashimoto, A.; Ikeshiro, Y.; Okabe, H.; Cosentino, L.M.; Lee, K.H. Anti-AIDS agents 38. Anti-HIV activity of 3-O-acyl ursolic acid derivatives. *J. Nat. Prod.* **2000**, *63*, 1619–1622. [CrossRef]
38. Yan, S.; Huang, C.; Wu, S.; Yin, M. Oleanolic acid and ursolic acid induce apoptosis in four human liver cancer cell lines. *Toxicol. In Vitro* **2010**, *24*, 842–848. [CrossRef]
39. Kowalski, R. Studies of selected plant raw materials as alternative sources of triterpenes of oleanolic and ursolic acid types. *J. Agric. Food Chem.* **2007**, *55*, 656–662. [CrossRef]
40. Zheljazkov, V.D.; Astatkie, T. Effect of residual distillation water of 15 plants and three plant hormones on Scotch spearmint (*Mentha* × *gracilis* Sole). *Ind. Crop. Prod.* **2011**, *33*, 704–709. [CrossRef]
41. Tiliacos, C.; Gaydou, E.M.; Bessiere, J.-M.; Agnel, R. Distilled lavandin (*Lavandula intermedia* Emeric ex. Loise l) wastes: A rich source of coumarin and herniarin. *J. Essent. Oil Res.* **2008**, *20*, 412–413. [CrossRef]
42. Torras-Claveria, L.; Jauregui, O.; Bastida, J.; Codina, C.; Viladomat, F. Antioxidant Activity and Phenolic Composition of Lavandin (*Lavandula* x *intermedia* Emeric ex. Loiseleur). *J. Agric. Food Chem.* **2007**, *55*, 8436–8443. [CrossRef]
43. Khadzhieva, P.; Aleksiev, K.; Topalova, I. Di-and triterpenoids in essential oil waste-from pinus, lavender, and salvia. *Int. Conf. Chem. Biotechnol. Biol. Act. Nat. Prod.* **1987**, *5*, 519–524.
44. Amorese, V.; Donadu, M.; Usai, D.; Sanna, A.; Milia, F.; Pisanu, F.; Molicotti, P.; Zanetti, S.; Doria, C. In vitro activity of essential oils against *Pseudomonas aeruginosa* isolated from infected hip implants. *J. Infect. Dev. Ctries.* **2018**, *12*, 996–1001. [CrossRef]
45. Yap, P.S.X.; Krishnan, T.; Yiap, B.C.; Hu, C.P.; Chan, K.-G.; Lim, S.H.E. Membrane disruption and anti-quorum sensing effects of synergistic interaction between *Lavandula angustifolia* (lavender oil) in combination with antibiotic against plasmid-conferred multi-drug-resistant *Escherichia coli*. *J. Appl. Microbiol.* **2014**, *116*, 1119–1128. [CrossRef]
46. Yap, P.S.X.; Lim, S.H.E.; Hu, C.P.; Yiap, B.C. Combination of essential oils and antibiotics reduce antibiotic resistance in plasmid-conferred multidrug resistant bacteria. *Phytomedicine* **2013**, *20*, 710–713. [CrossRef] [PubMed]
47. Kunicka-Styczyńska, A.; Śmigielski, K.; Prusinowska, R.; Rajkowska, K.; Kuśmider, B.; Sikora, M. Preservative activity of lavender hydrosols in moisturizing body gels. *Lett. Appl. Microbiol.* **2015**, *60*, 27–32. [CrossRef]
48. *ISO 3515:2002 Oil of Lavender (Lavandula Angustifolia Mill.)*; ISO: Geneva, Switzerland, 2002.
49. Ivanov, I.; Petkova, N.; Tumbarski, Y.; Vrancheva, R.; Stoyanova, M. Lavender waste–promising source of triterpenoids and polyphenols with antioxidant and antimicrobial activity. *Ind. Technol.* **2018**, *5*, 26–32.
50. Piqué, N.; Miñana-Galbis, D.; Merino, S.M.; Tomás, J. Virulence Factors of *Erwinia amylovora*: A Review. *Int. J. Mol. Sci.* **2015**, *16*, 12836–12854. [CrossRef] [PubMed]
51. Benada, M.; Boumaaza, B.; Boudalia, S.; Khaladi, O.; Guessas, B. Variability of aggressiveness and virulence of *Erwinia carotovora* subsp. *carotovorum* causing the soft rot on potato tubers in the western of Algeria. *Int. J. Plant Biol.* **2018**, *9*, 52–56. [CrossRef]
52. Tamir-Ariel, D.; Navon, N.; Burdman, S. Identification of Genes in *Xanthomonas campestris* pv. *vesicatoria* induced during its interaction with tomato. *J. Bacteriol.* **2007**, *189*, 6359–6371. [PubMed]
53. Davies, J.C. *Statistics and Data Analysis in Geology*; Elsevier: Amsterdam, The Netherlands, 2002.
54. Shi, G. Cluster Analysis. In *Data Mining and Knowledge Discovery for Geoscientists*; Shi, G., Ed.; Elsevier: Amsterdam, The Netherlands, 2014.
55. Ciocarlan, A.; Aricu, A.; Geană, E.-I. Investigating the therapeutic potential of some medicinal plant and wild fruits based on terpenic ursolic and oleanolic acids. *Cient Period. Nutr.* **2019**, *3*, 1–12.
56. Geană, E.I.; Ionete, R.E.; Ciocarlan, A.; Aricu, A.; Fulga, A.; Ungur, N.; Podogova, M.; Nikolaeva, D. HPLC determination of oleanolic and ursolic acids in apples and by-products. *Prog. Cryog. Isot. Separation* **2014**, *17*, 53–62.
57. Lungu, L.; Ciocarlan, A.; Barba, A.; Shova, S.; Pogrebnoi, S.; Mangalagiu, I.; Moldoveanu, C.; Vornicu, N.; D'Ambrosio, M.; Babak, M.V.; et al. Synthesis and evaluation of biological activity of homodrimane sesquiterpenoids bearing hydrazinecarbothioamide or 1,2,4-triazole unit. *Chem. Heter. Comp.* **2019**, *55*, 716–724. [CrossRef]
58. Lupașcu, G.; Ciocarlan, A.; Dragalin, I.; Lupașcu, L. Antimicrobial activity of the Coriander oil (*Coriandrum sativum* L.). *Rom. J. Biol.* **2019**, *64*, 31–42.

plants

MDPI

Article

Biomolecular Evaluation of *Lavandula stoechas* L. for Nootropic Activity

Aamir Mushtaq [1,2], Rukhsana Anwar [1], Umar Farooq Gohar [3], Mobasher Ahmad [1,2], Romina Alina Marc (Vlaic) [4,*], Crina Carmen Mureşan [4], Marius Irimie [5,*] and Elena Bobescu [5]

1 Department of Pharmacology, Punjab University College of Pharmacy, University of the Punjab, Lahore 54000, Pakistan; aamir_mushtaq@hotmail.com (A.M.); rukhsanaanwar2003@yahoo.com (R.A.); ahmadmobasher@hotmail.com (M.A.)
2 Gulab Devi Institute of Pharmacy, Gulab Devi Educational Complex, Lahore 54000, Pakistan
3 Institute of Industrial Biotechnology, GC University, Lahore 54000, Pakistan; dr.mufgohar@gcu.edu.pk
4 Food Engineering Department, Faculty of Food Science and Technology, University of Agricultural Sciences and Veterinary Medicine, 400372 Cluj-Napoca, Romania; crina.muresan@usamvcluj.ro
5 Faculty of Medicine, Transilvania University of Brasov, 500036 Brasov, Romania; elena.bobescu@gmail.com
* Correspondence: romina.vlaic@usamvcluj.ro (R.A.M.); marius_irimie2002@yahoo.com (M.I.); Tel.: +40-740379436 (R.A.M. & M.I.)

Abstract: *Lavandula Stoechas* L. is widely known for its pharmacological properties. This study was performed to identify its biomolecules, which are responsible for enhancement of memory. *L. stoechas* aqueous extract was first purified by liquid column chromatography. The purified fractions were analyzed for in vitro anti-cholinesterase activity. The fraction that produced the best anti-cholinesterase activity was named an active fraction of *L. stoechas* (AfL.s). This was then subjected to GC–MS for identifications of biomolecules present in it. GC–MS indicated the presence of phenethylamine and α-tocopherol in AfL.s. Different doses of AfL.s were orally administered (for seven days) to scopolamine-induced hyper-amnesic albino mice and then behavioral studies were performed on mice for two days. After that, animals were sacrificed and their brains were isolated to perform the biochemical assay. Results of behavioral studies indicated that AfL.s improved the inflexion ratio in mice, which indicated improvement in retention behavior. Similarly, AfL.s significantly ($p < 0.001$) reduced acetylcholinesterase and malondialdehyde contents of mice brain, but on the other hand, it improved the level of choline acetyltransferase, catalase, superoxide dismutase, and glutathione. It was found that that high doses of AfL.s (≥400 mg/Kg/p.o.) produced hyper-activity, hyperstimulation, ataxia, seizures, and ultimate death in mice. Its LD_{50} was calculated as 325 mg/Kg/p.o. The study concludes that α-tocopherol and phenethylamine (a primary amine) present in *L. stoechas* enhance memory in animal models.

Keywords: phenethylamine; *L. stoechas*; acetylcholine; choline acetyltransferase; AChE; aromatic amine; enhancement of memory

Citation: Mushtaq, A.; Anwar, R.; Gohar, U.F.; Ahmad, M.; Marc (Vlaic), R.A.; Mureşan, C.C.; Irimie, M.; Bobescu, E. Biomolecular Evaluation of *Lavandula stoechas* L. for Nootropic Activity. *Plants* **2021**, *10*, 1259. https://doi.org/10.3390/plants10061259

Academic Editor: Antonella Smeriglio

Received: 30 April 2021
Accepted: 16 June 2021
Published: 21 June 2021

1. Introduction

Lavandula stoechas L. (Lamiaceae) is an aromatic and medicinal plant of the Mediterranean region. It was the most popular folk remedy for the management of digestive disorders, kidney disease, diabetes mellitus, hyperlipidemia, cough, asthma, headache, and flu-like symptoms [1,2]. It is also known as "broom of the brain" due to its extensive use in the treatment of migraines, epilepsy, and memory related disorders [3,4]. The plant is rich in camphor, erythrodiol, eucalyptol, fenchone, *lavanol*, longipene-2-ene, longipene-2-ene monoacetate, lupeol, luteolin, myrtenol, oleanolic acid, pinocarvyl acetate, terpineol, ursolic acid, vergatic acid, vitexin, α-amyrin, β-sitosterol, and a variety of aromatic compounds [5,6]. Linalool present in *L. stoechas* has a sedative effect [7] and it acts on different brain receptors to modify behavioral patterns, along with creating a sense of calmness and wellbeing. Camphor, present in *L. stoechas,* has stimulating effects on the brain [8]. Lavender oils extracted from *L. stoechas* are famous for their spicy fragrances and are used in aromatherapy to treat anxiety, depression, sleeplessness, headaches, and migraine [9]. Moreover, *L. stoechas* showed antispasmodic [8], sedative, anti-epileptic, antibacterial [10], antifungal [11], anti-leishmaniasis [5], anti-inflammatory [12], cytotoxic [13] and anti-diabetic [14] properties.

Keeping in mind the traditional uses of *L. stoechas*, as a neurotonic and memory enhancer, as part of our studies on the methanolic extract of *L. stoechas* [15], our current study was designed to report the active ingredient of *L. stoechas* aqueous extract responsible for nootropic activity.

2. Materials and Methods

2.1. Chemicals and Drugs

The following chemicals used were of analytical grade, and procured from Sigma Aldrich, MS Traders, Lahore Pakistan: 2,2-*diphenyl*-1-picrylhydrazyl (DPPH) (95%), 5,5′-dithiobis-2-nitrobenzoic acid (*DTNB*) (99%), 4,4-dithiodipyridine (98%), Folin–Ciocalteu reagent (FCR), trichloroacetic acid (TCA) (98%), thiobarbituric acid (TBA) (99%), tannic acid (99.9%), superoxide dismutase (SOD), sodium dodecyl sulfate, sodium carbonate (99%), rutin, reduced nicotinamide adenine dinucleotide (NADH), potassium dichromate (99.5%), potassium acetate (99%), phenyl methanesulfonate (PMS), nitro blue tetrazolium (NBT), *n*-butanol (99.8%), *n*-hexane (99%), methanol (99.8%), hydrogen peroxide (H_2O_2) (85%), gallic acid, ethanol, chloroform (99%), carboxy methyl cellulose (CMC), ascorbic acid (99%), aluminum chloride (99.9%), acetyl thiocholine iodide, silica gel-60, acetyl coenzyme-A, acetylcholine esterase, choline chloride (98%), EDTA, fast blue B salt, neostigmine sulfate, β-naphthyl acetate, and acetic acid. Piracetam was gifted from Jiangxi Yuehua Pharmaceutical, China, and scopolamine was obtained from Merck Pharmaceutical Pvt. Ltd. (Kenilworth, NJ, USA).

2.2. Extraction and Fractionation by Column Chromatography

L. stoechas L. (aerial parts) was purchased from a local market in Lahore, Pakistan. It was then identified by a botanist from the department of Botany, Government College University (GCU), Lahore. The specimen was preserved in the herbarium of the GCU and was assigned a voucher number: GC.Herb.Bot.3386. The extraction was conducted in methanol, by using simple maceration, as described in our previous study [15]. Then, fractional extraction was conducted by using different solvents, i.e. *n*-hexane, chloroform, ethyl acetate, and *n*-butanol. Finally, the aqueous layer left behind was obtained and the solvent was evaporated to obtain a semi-solid aqueous extract of *L. stoechas*. Column chromatography was used for the fractionation of the aqueous extract of *L. stoechas*. An appropriate-sized glass column was packed with slurry, which was made by dissolving almost 10 g of silica gel-60 in the same solvent as used for the mobile phase. Standard protocols were followed for the packing and running of the column. *L. stoechas* aqueous extract was dissolved in a small amount of solvent; that sample mixture was loaded in a column via a pipette. The space above the sample in the column was filled with solvent and the stopper was opened to obtain the separated fraction in a flask below the outlet of the column [16]. This way, different solvent systems were used for the separation of different constituents in the crude extract. Separated fractions were collected in the test tubes and labeled for further tests.

2.3. Anti-Cholinesterase Activity (In Vitro Assay)

In vitro anti-cholinesterase activity was performed on all fractions obtained through column chromatography by using an (NA-FB) microwell plate assay. The solution was prepared by dissolving β-naphthyl acetate (0.25 mg) in methanol (1 mL); 50 μL of this was mixed with 10 μL of plant extract. Then, 200 μL of acetylcholine esterase solution (3.33 U/mL) was poured in the reaction mixture by keeping the temperature of the mixture at 4 °C. It was incubated for 40 min at the same temperature and then 2.5 mg of fast blue B salt was dissolved in 1 mL of distilled water. Out of which, 10 µL was dropped into the incubated reaction mixture; a change in the solution color was observed. β-naphthyl acetate was used as a substrate, while fast blue B salt was used as a color reagent. The principle applies that β-naphthyl acetate is hydrolyzed into naphthol acetic acid by AChE. Naphthol then reacts with fast blue B salt, imparting a purple color to the mixture. No change in color of the solution indicates a strong anti-cholinesterase activity of the fraction, while a dark purple color indicates no inhibition of AChE [17].

2.4. GC–MS Analysis

The bioactive compounds present in the final fraction of *L. stoechas* were detected by performing gas chromatography mass spectroscopy (GC–MS) analysis by using GC–MS equipment (Agilent 6890N). The TR-5-MS capillary non-polar standard column, with dimensions of 30 Mts, an internal diameter of 0.25 mm, and film thickness of 0.25 µm, was used. Helium gas (99.99%) was used as carrier gas and mobile phase was run with a flow rate of 1 mL/min. The starting temperature of the oven was 40 °C, which was raised to 250 °C @ 10 °C/min. The sample was dissolved in methanol and an aliquot of 2 µL was injected by keeping the temperature of the injector and detector fixed at 250 °C and 280 °C, respectively, while the ion source temperature was fixed at 200 °C. Identification of the compound was by molecular mass and the structure of the compound by interpretation of the GC–MS standard library.

2.5. Behavioral and Biochemical Studies

Behavioral studies were performed by using elevated plus maze, light/dark test, and hole-board test models, using standard protocols. Biochemical studies were performed to assess the level of acetylcholinesterase (AChE), malondialdehyde (MDA), superoxide dismutase (SOD), catalase (CAT), and glutathione (GSH) in the brain homogenates of mice, of all groups, using standard protocols. Detailed procedures of these tests are described in a previously published paper in this series [15].

2.6. Animals

All studies were performed on male Swiss albino mice, which were provided standard living conditions, as described in a previous manuscript [15], while female albino mice (20–25 g) were used for toxicity studies.

2.7. Study Design for Behavioral/Biochemical Studies

For behavioral/biochemical studies, mice were divided into seven groups (n = 6) and were treated accordingly, as shown in Table 1. While ChAT was performed on five groups (n = 6) of mice, which were treated accordingly, as shown in Table 2.

Table 1. Study design for behavioral and biochemical studies.

Groups	Treatment from Days 1–7	Treatment on Day 7, 45 min after Administration of the Last Dose
G-I (Normal Control)	Normal saline 10 mL/Kg/p.o.	------
G-II (Amnesic Control)	5% CMC 10 mL/Kg/p.o.	Scopolamine (10 mg/Kg/p.o.)
G-III (Standard Control-A)	Piracetam 200 mg/Kg/p.o.	------
G-IV (Standard Control-B)	Piracetam 200 mg/Kg/p.o.	Scopolamine (10 mg/Kg/p.o.)
G-V (Experimental Control-I)	AfL.s 9 mg/Kg/p.o.	Scopolamine (10 mg/Kg/p.o.)
G-VI (Experimental Control-II)	AfL.s 18 mg/Kg/p.o.	Scopolamine (10 mg/Kg/p.o.)
G-VII (Experimental Control-III)	AfL.s 18 mg/Kg/p.o.	------

Doses were prepared by suspending AfL.s in CMC (5%) and by dissolving piracetam and scopolamine in normal saline. Then, behavioral studies were performed on days 7 and 8, and animals were sacrificed for the performance of biochemical studies on the eigth day after completing behavioral trials.

Table 2. Study design for the assessment of choline acetyltransferase (ChAT) activity.

Groups	Treatment		
	Day 1	Days 2–6	Day 7
G-I	Normal Saline (10 mL/Kg/p.o.)	Normal Saline (10 mL/Kg/p.o.)	Normal Saline (10 mL/Kg/p.o.)
G-II	5% CMC (10 mL/Kg/p.o.)	5% CMC (10 mL/Kg/p.o.)	Scopolamine (10 mg/Kg/P.O)
G-III	AfL.s (18 mg/Kg/p.o.)	AfL.s (18 mg/Kg/p.o.)	AfL.s (18 mg/Kg/p.o.)
G-IV	Scopolamine (10 mg/Kg/p.o.)	AfL.s (18 mg/Kg/p.o.)	AfL.s (18 mg/Kg/p.o.)
G-V	AfL.s (18 mg/Kg/p.o.)	AfL.s (18 mg/Kg/p.o.)	AfL.s (18 mg/Kg/p.o.) + Scopolamine (10 mg/Kg/p.o.)

Two hours after administration of the last dose, the animals were sacrificed by using chloroform to get their brain. Then brain homogenates were formed according to standard procedure [15] and ChAT activity was determined by the above-described spectroscopic method.

2.8. Choline Acetyltransferase Activity (ChAT)

The reagent was prepared by mixing 10 µL of each sodium phosphate buffer (0.5 M, pH 7.2), sodium chloride (3 M), neostigmine sulfate (7.6×10^{-4} M), acetyl coenzyme-A (6.2×10^{-3} M prepared in 0.01N HCl), EDTA (1.1×10^{-3} M), choline chloride (1 M), and sodium chloride (3 M), and incubated at 37 °C for 5–10 min. Brain homogenate (100 µL) was mixed in the smallest volume of the reagent and the final volume was 0.2 mL. It was then boiled in a water bath for 2 min after incubating at 37 °C for 25 min. Then, oxygen-free distilled water was added to the mixture, which was then centrifuged at high speed to separate out the denatured protein. Then, 10 µL of 4,4-dithiodipyridine (10^{-3} M) was added to 0.5 mL of the supernatant and absorbance was read at 324 nm against the blank in a UV-visible spectrophotometer [18].

2.9. Acute Toxicity Studies

Acute toxicity study was performed on female albino mice according to OECD guidelines 423 2001. Initially, the pilot study was done on a small number of mice to find the dose range at which animals started to die. It was found that active fractions of *L. stoechas* (AfL.s) showed no death up to 300 mg/Kg/p.o. and showed 100% mortality at 500 mg/Kg/p.o. Thus, animals were divided into six groups ($n = 5$). Group I was normal control; animals in Groups II to VI were administered with AfL.s in wide, spaced doses, in ascending order (300, 350, 400, 450, and 500 mg/Kg/p.o., respectively). They were then observed for 24 h to find the number of mortalities. Finally, LD_{50} was calculated by using the following formula:

$$LD_{50} = \text{Least Lethal Dose} - \Sigma (a \times b)/n$$

The animals who survived in different groups after administration of acute toxic dose were observed for two weeks for the assessment of physical and behavioral changes. Ataxia, blanching, convulsions, cyanosis, depression, hyperactivity, hypnosis, irritability, jumping, loss of traction, muscle spasm, piloerection, redness, rigidity, salivation, secretions, sedation, stimulation and straub reaction were the parameters that were observed after acute doses administration.

2.10. Statistical Analysis

The data are expressed as mean ± SEM. Student's *t*-test analysis was applied on data with paired comparisons, and multiple comparisons were made by ANOVA followed by Dunnett's test by using GraphPad Prism software version 7. Value of $p < 0.05$ was marked as significant.

3. Results

3.1. Fractionation by Column Chromatography

L. stoechas aqueous extract was fractionated further by column chromatography. In total, 55 fractions were obtained by using different solvent systems (based on polarity).

The fractions were evaluated by thin layer chromatography (TLC), were combined, and again passed from the column; finally, there were 15 fractions (Figure 1).

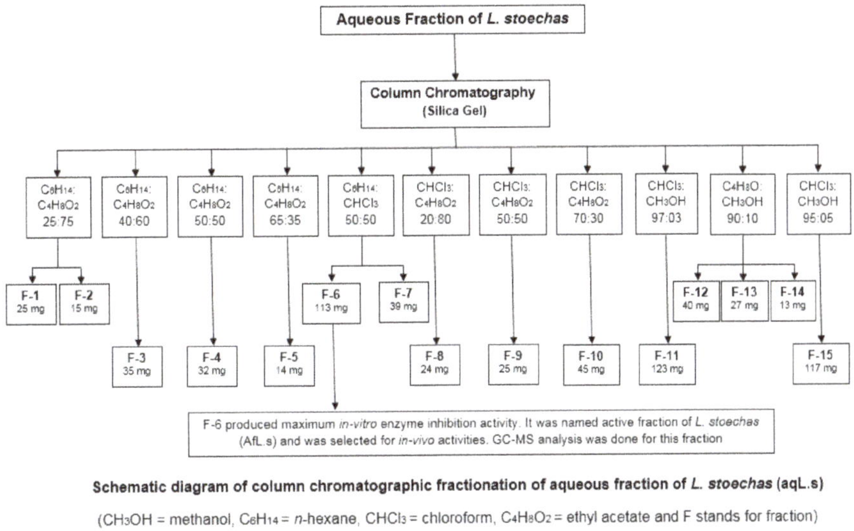

Figure 1. General scheme of fractionation by column chromatography.

Anti-Cholinesterase Activity (In Vitro)

In vitro testing indicated that fraction no. 6 showed the best enzyme inhibition among all the fractions, as expressed in Table 3. Fraction no. 6 was named an active fraction of *L. stoechas* (AfL.s) and was tested for chemical analysis by GC–MS. Furthermore, in-vivo studies (behavioral and biochemical) were performed on mice to find the memory enhancing effects of AfL.s.

Table 3. Anti-cholinesterase activity (in vitro) shown by different fractions of *L. stoechas*.

No.	Fractions	Color of Solution	Inhibition of AChE
1	F-1	Dark purple	No
2	F-2	Dark purple	No
3	F-3	Dark purple	No
4	F-4	Dark purple	No
5	F-5	Dark purple	No
6	F-6	No color change	Very strong
7	F-7	Light purple	Mild
8	F-8	Dark purple	No
9	F-9	Dark purple	No
10	F-10	Dark purple	No
11	F-11	Dark purple	No
12	F-12	Light purple	Mild
13	F-13	Light purple	Mild
14	F-14	Light purple	Mild
15	F-15	Dark purple	No

Fraction no. 6 (F-6) possessed strong anti-cholinesterase activity, so it was selected for further chemical and in-vivo studies.

3.2. GC–MS Analysis of AfL.s

GC–MS analysis indicated that AfL.s contained the phenethylamine group of compounds and cholestan-7-one. The details are shown in Table 4 and spectrum is given in Figure S1.

Table 4. Compounds detected in AfL.s by GC–MS.

	Compound Name	Molecular Formula	Molecular Weight (g/mol)	Mass Peak	Retention Time (min)
1	Phenethylamine, N methyl-beta 3,4 (trimethylsiloxy)	$C_{18}H_{37}NO_3Si_3$	399	50	21.167
2	Cholestan-7-one	$C_{29}H_{50}O_2$	430	56	24.083
3	Phenethylamine	$C_{18}H_{37}NO_3Si_3$	399	50	21.167
4	N-Methyladrenaline	$C_{19}H_{39}NO_3Si_3$	413	50	21.167
5	Benzeneacetic acid	$C_{20}H_{42}O_5Si_4$	472	50	21.167

3.3. Behavioral Studies (Effect of AfL.s on EPM, Light/Dark Test, and Hole-Board Paradigm in Mice)

Results of behavioral studies indicated that animals treated with AfL.s significantly ($p < 0.001$) reduced the initial transfer latencies (ITL) and retention transfer latencies (RTL) in comparison to the amnesic control group (Figure 2). Similarly, inflexion ratio (IR) calculated from ITL and RTL values, indicated that active fraction-treated mice showed maximum IR value (0.17 ± 0.04) in comparison to scopolamine-treated mice (−0.19 ± 0.04). Thus, higher IR values indicated significant ($p < 0.001$) improvement of memory in AfL.s-treated mice.

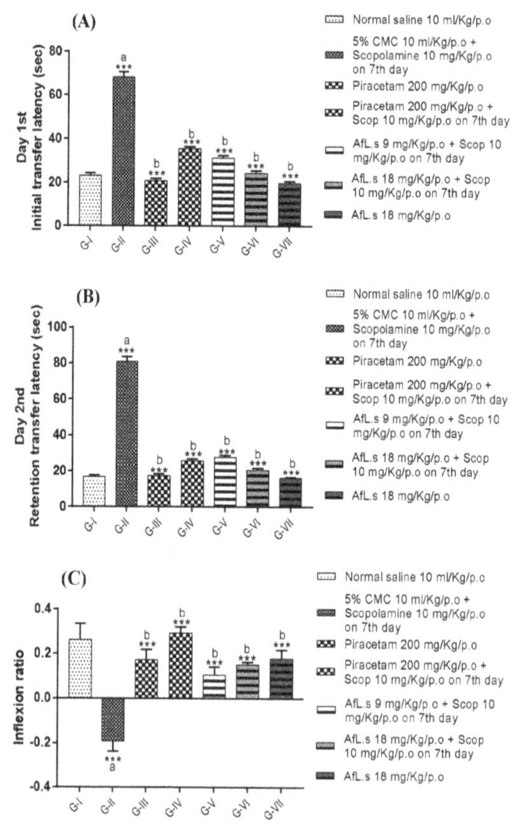

Figure 2. *Cont.*

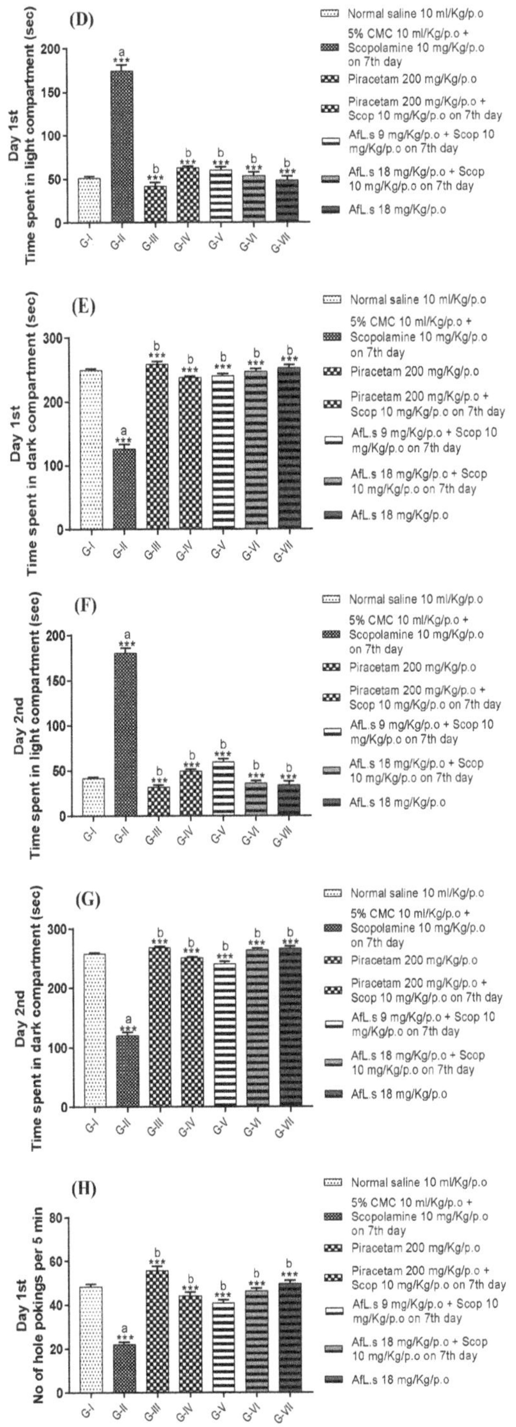

Figure 2. *Cont.*

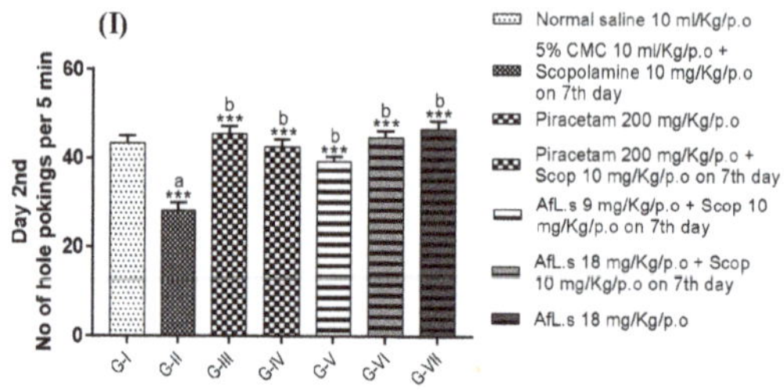

Figure 2. Effect of AfL.s on (**A**) initial transfer latency; (**B**) retention transfer latency; (**C**) inflexion ratio in the elevated plus maze paradigm; (**D**) time spent (sec) in the light compartment on day 1; (**E**) time spent (sec) in the dark compartment on day 1; (**F**) time spent (sec) in the light compartment on day 2; (**G**) time spent (sec) in the dark compartment on day 2; (**H**) number of hole pokings by mice on day 1, and (**I**) number of hole pokings by mice on day 2. Data are presented as mean ± SEM (n = 6) and one-way ANOVA (Dunnett's test) was applied by comparing G-II to G-I (presented by "a" on the bar). All other groups were compared to G-II (presented by "b" on the bar). The signs *ns*, *, ** and *** presented the p values as ≥0.05, ≤0.05, ≤0.01, and ≤0.001, respectively).

Results of the light/dark test indicated that AfL.s-treated mice spent most of the time in the dark portion of the apparatus, on the first and second day of observation, in comparison to the amnesic control group. This finding is based on the principle that animals in the amnesic control group forgot to find the dark area of the apparatus while the standard control and AfL.s-treated mice retained their memory of exploration. Thus, it is clear that AfL.s significantly ($p < 0.001$) improved the memory in AfL.s-treated mice (Figure 2).

Similarly, results of hole-board paradigm indicated that amnesic control animals significantly reduced the number of hole-pokings while standard control and AfL.s-treated mice retained their memory of exploration and showed a significantly ($p < 0.001$) increased number of hole-pokings on both the first and second day of observation (Figure 2).

3.4. Biochemical Studies (Effect of AfL.s on Levels of AChE, MDA, SOD, CAT, and GSH in Mice Brains)

Biochemical studies indicated that the level of acetylcholinesterase (AChE) was significantly ($p < 0.001$) reduced in group-II (scopolamine treated) animals, while group-VII animals (AfL.s 18 mg/Kg/p.o.) showed maximum inhibition of AChE among all the groups (Figure 3A). Similarly, the level of MDA is reduced significantly ($p < 0.001$) in AfL.s-treated mice as compared to the amnesic control group (which showed the highest MDA level) (Figure 3B). It was observed that the levels of SOD, CAT, and GSH increased significantly in group-V (AfL.s 9 mg/Kg/p.o.) among all groups (Figure 3C–E). This observation leads to the fact that AfL.s possesses strong antioxidant activity when used in low doses.

(**A**)

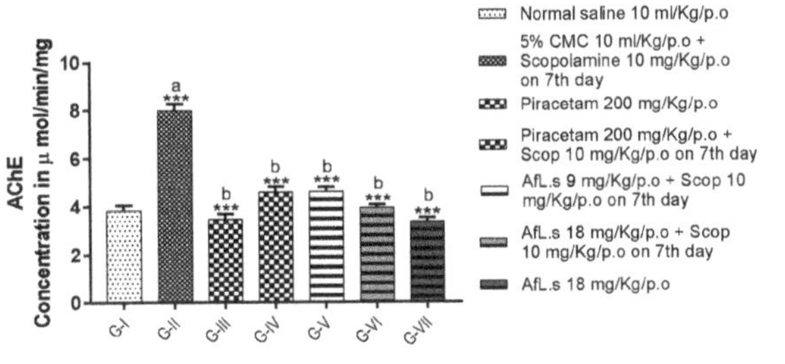

(**B**)

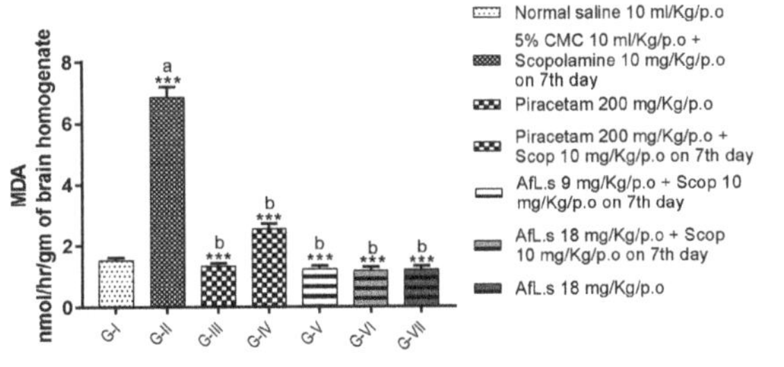

(**C**)

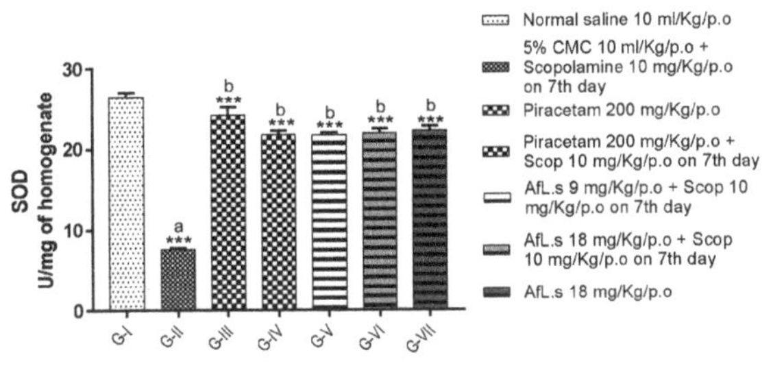

(**D**)

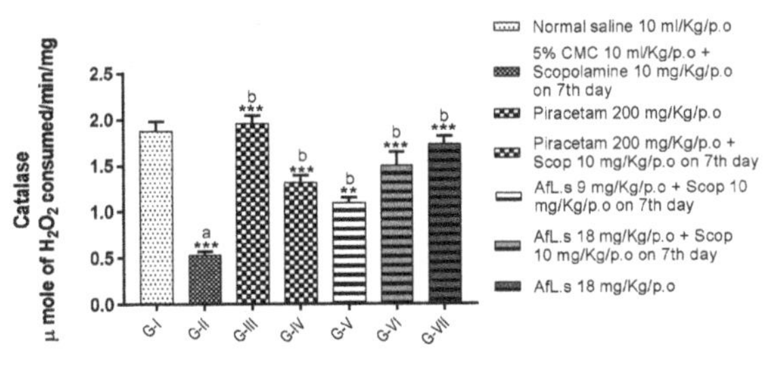

Figure 3. *Cont.*

(E)

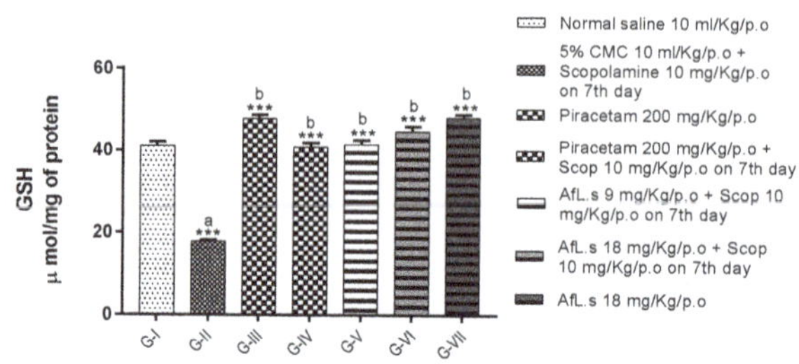

Figure 3. Effect of AfL.s on concentration of (**A**) acetylcholinesterase (AChE); (**B**) MDA; (**C**) SOD; (**D**) CAT; and (**E**) GSH in brain homogenate. Data are presented as mean ± SEM ($n = 6$) and one-way ANOVA (Dunnett's test) was applied by comparing G-II to G-I (presented by "a" on bar). All other groups were compared to G-II (presented by "b" on bar). The signs *ns*, *, ** and *** presented the p values as ≥0.05, ≤0.05, ≤0.01, and ≤0.001, respectively).

3.5. Effect of AfL.s on ChAT Activity

The level of ChAT was observed, 11.85 ± 0.92, 7.59 ± 0.76, 18.13 ± 1.23, 16.70 ± 1.16, 12.10 ± 1.45, 11.63 ± 0.66, 10.84 ± 1.22, and 7.08 ± 0.68 µmol/min/mg, from group-I to V, respectively. Thus it is clear that animals treated only with AfL.s (18 mg/Kg/p.o.) for seven consecutive days showed the best elevation in ChAT levels as compared to the normal control group, with the level of significance as ($p < 0.01$), as shown in Figure 4.

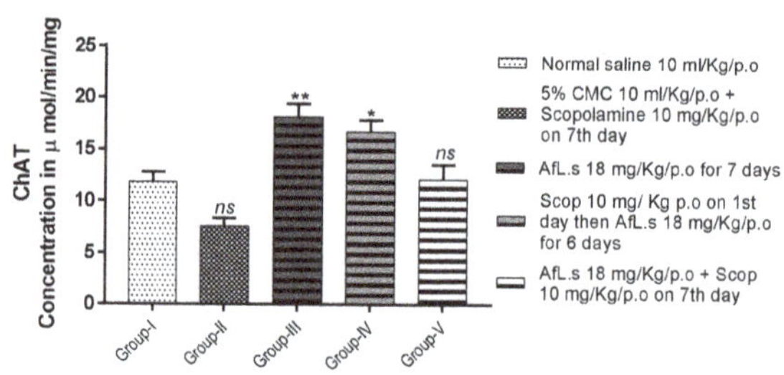

Figure 4. Effect of AfL.s on concentration of ChAT in mice brains. Data are presented as mean ± SEM ($n = 6$) and one-way ANOVA (Dunnett's test) was applied by comparing all groups with G-I. The signs *ns*, *, ** and *** presented the p values as ≥ 0.05, ≤ 0.05, ≤ 0.01, and ≤ 0.001 respectively.

3.6. Acute Toxicity Study

Acute toxicity study performed on six groups ($n = 5$) of female albino mice indicated that AfL.s produced no death up to a dose of 350 mg/Kg/p.o. Animals began to die when the dose increased above 350 mg/Kg/p.o. All the animals died when they were administered with a single oral dose 500 mg/Kg of AfL.s. All the details and calculations for median lethal dose LD_{50} are provided in (Table 5). LD_{50} for AfL.s was calculated as 325 mg/Kg/p.o. The animals who survived after the administration of an acute single dose of AfL.s 450 mg/Kg/p.o. were kept in observations for two weeks for the assessment of behavioral and physiological changes. It was observed that animals positively exhibited stimulation, straub reaction. salivation, rigidity, piloerection, other secretions, muscle spasm, loss of traction, jumping, irritability, hyperactivity, convulsions, blanching, and ataxia, while, sedation, redness, ptosis, hypnosis, depression, and cyanosis were not observed in the mice (Table 6).

Table 5. Calculation of median lethal dose LD_{50} of AfL.s.

Groups	Dose Difference (a)	Mortality	Mean Mortality (b)	(a × b)
G-I (Normal Control)	0	0	0	0
G-II (AfL.s 300 mg/Kg/p.o.)	300	0	0	0
G-III (AfL.s 350 mg/Kg/p.o.)	50	0	0	0
G-IV (AfL.s 400 mg/Kg/p.o.)	50	2	2 + 0/2 = 1	50
G-V (AfL.s 450 mg/Kg/p.o.)	50	3	3 + 2/2 = 2.5	125
G-VI (AfL.s 500 mg/Kg/p.o.)	50	5	5 + 3/2 = 4	200
				Σ (a × b) = 375

$\Sigma(a \times b)/n = 375/5$; LD_{50} = Least lethal Dose − $\Sigma(a \times b)/n$; 400 − 375/5 = 325 mg/Kg.

Table 6. Effect of acute toxic dose of AfL.s 450 mg/Kg/p.o. on the behavior and physiology of mice.

Behavioral Changes	Days													
	1	2	3	4	5	6	7	8	9	10	11	12	13	14
Straub Reaction	-	-	+	+	+	+	+	+	+	-	-	-	-	-
Stimulation	+	+	+	+	+	+	+	+	+	+	-	-	-	-
Sedation	-	-	-	-	-	-	-	-	-	-	-	-	-	-
Secretions	+	+	+	+	+	+	+	-	-	-	-	-	-	-
Salivation	+	+	+	+	+	-	-	-	-	-	-	-	-	-
Rigidity	+	+	+	+	+	+	+	-	-	-	-	-	-	-
Redness	-	-	-	-	-	-	-	-	-	-	-	-	-	-
Ptosis	-	-	-	-	-	-	-	-	-	-	-	-	-	-
Piloerection	+	-	-	-	-	-	-	-	-	-	-	-	-	-
Muscle Spasm	+	+	+	+	-	-	-	-	-	-	-	-	-	-
Loss of Traction	-	-	-	-	-	-	+	+	+	+	+	+	+	+
Jumping	+	+	+	+	+	+	+	+	+	+	+	-	-	-
Irritability	+	+	+	+	+	+	+	+	+	+	+	+	+	+
Hypnosis	-	-	-	-	-	-	-	-	-	-	-	-	-	-
Hyperactivity	+	+	+	+	+	+	+	+	+	+	+	+	+	+
Depression	-	-	-	-	-	-	-	-	-	-	-	-	-	-
Cyanosis	-	-	-	-	-	-	-	-	-	-	-	-	-	-
Convulsions	+	+	+	-	-	-	-	-	-	-	-	-	-	-
Blanching	+	+	+	+	+	+	-	-	-	-	-	-	-	-
Ataxia	+	+	+	+	+	+	+	+	-	-	-	-	-	-

"-"Sign indicates absence of effect while "+" sign indicates the presence of effect.

4. Discussion

Aromatic plants of the Mediterranean region have advanced medicinal values. Alcoholic and hydro-alcoholic extracts of many aromatic plants of Asteraceae, Apiaceae, and Lamiaceae families have been used extensively for their wide variety of pharmacological activities. Genus Lavandula contains as much as 39 species, including two economically well renowned species—*L. stoechas* and *L. angustifolia*. Hydroalcoholic extracts of *L. stoechas* contain catechic tannins, flavonoids, coumarins, sterols, mucilages, and leucoanthocyanins [19]. Different compounds, such as α-amyrin, β-sitosterol, α-amyrin acetate, oleanolic acid, vergatic acid, ursolic acid, erythrodiol, lupeol, luteolin, vitexin, acacetin, lavanol, 7-methoxy coumarin, and two longipinane derivatives (longipin-2-ene-7β,9α-diol-1-one-9-monoacetate and longipin-2-ene-7β,9α-diol-1-one) have been isolated from areal parts of *L. stoechas* [20]. *L. stoechas* exhibits a wide array of pharmacological activities; research shows that it boosts memory in albino mice [15]. This is an original research work specifically designed to report the biomolecules of *L. stoechas*, which boosts memory in mice brains. In this study, the *L. stoechas* aqueous extract was purified, and chemical characterization of AfL.s by GC–MS indicated the presence of two main compounds (phenethylamine and α-tocopherol) (Figure S1, Table 4). Past studies have scientifically proven the anticholinesterase activity of phenethylamine [21] and the antioxidant potential of cholestan-7-one, usually named α-tocopherol [22].

Phenethylamine has been scientifically proven as a brain neuromodulator and a strong inhibitor of AChE [23]. Phenethylamine is composed of an aromatic ring to which a side chain of two carbons having amine at the terminal position is attached. Substitution of alkyl groups at different positions on the phenyl ring would attribute to its strong neuromodulative and psychoactive activities. Thus, overall cognition and brain performance is enhanced by phenethylamine [21,24].

The results of the behavioral studies concluded that AfL.s significantly ($p < 0.001$) enhanced the retention power and learning capacity of the mice brains. Similarly, treatment of animal with AfL.s showed significant ($p < 0.001$) reduction in the level of AChE (Figure 3A). Inhibiting AChE improves cholinergic transmission and relieves the patient of memory loss [25].

On the other hand, the level of choline acetyltransferase (ChAT) was elevated (Figure 4) in mice brains. Both of these findings strongly suggest that the level of acetylcholine (ACh) increased in mice brains by dual mechanisms. This effect would be due to the action of phenethylamine on trace amine-associated receptors (TAARs), which are abundantly present in the brain, pituitary glands, kidney, liver, and stomach [26]. It has been proposed that phenethylamine binds with G-protein (either G_s or G_q subunit) coupled receptors (TAARs) and enhances memory and cognition [27,28].

The possible antioxidant mechanism of AfL.s is due to the presence of α-tocopherol, which not only promotes the glutathione level in the brain, but also causes attenuation of reactive oxygen species [22]. Furthermore, α-tocopherol is responsible for the formation of the stable and inert tocopheroxyl radical, by reacting with lipid peroxyl radical (LOO˙), as shown in Equation [29].

$$\text{LOO}^{\cdot} + \alpha\text{-tocopherol-OH} \rightarrow \text{LOOH} + \alpha\text{-tocopherol-O}^{\cdot}$$

Loss of memory may take place, either due to reduction in the level of acetylcholine [30] or by deposition of the β-amyloid protein [31] in the cerebral cortex and hippocampus of the brain [32]. Moreover, severe oxidative damage to neuronal circuits in the brain is another leading cause of memory loss [33,34], which is exhibited in scopolamine-treated mice [35]. The findings of the current study indicate that scopolamine-treated mice showed marked elevation of AChE (Figure 3A) and reduction in ChAT (Figure 4). Oxidative stress induced by scopolamine is the main factor behind this enzyme disturbance [36]. The treatment of mice with AfL.s not only reduced the level of AChE (Figure 3A), but also significantly ($p < 0.001$) boosted the level of ChAT (Figure 4) in mice brains. It is proposed that elevation in ChAT levels by AfL.s is caused by the antioxidant action of α-tocopherol on the brain. Similarly, phenethylamine present in AfL.s is responsible for the enzyme-mediated release of ACh in the brain [37]. The results also clearly indicate that animals only treated with AfL.s, without prior or subsequent administration of scopolamine, produced the highest elevation of ChAT levels in the brain (Figure 4).

Acute toxicity study indicated that AfL.s is toxic when used in high doses (≥400 mg/Kg/p.o.), which produces hyperactivity, hyperstimulation, ataxia, seizures, and ultimate death (Table 6). This toxicity is due to the primary toxic effect of high doses of phenethylamine, which is responsible for headaches, confusions, hallucination, seizures, and ultimately death in human beings [38]. Restlessness, diarrhea, headache, aggression, and tremors are mild side effects, which may be observed with the overdoses of phenethylamine [39].

Past studies have reported that high toxic doses (125–200 mg/Kg/i.p.) of phenethylamine produced very severe seizures and ultimate deaths due to cardiac arrest and overstimulation of the brain [40]. The LD_{50} for AfL.s was calculated as 325 mg/Kg/p.o. (Table 5), which indicated that it had a broad therapeutic index.

5. Conclusions

Two principle compounds—α-tocopherol and phenethylamine, present in *L. stoechas*—are responsible for the attenuation of dementia. α-tocopherol reduces oxidative stress of the free radicals in mice brains while phenethylamine enhances the level of acetylcholine in the hippocampus of mice brains. Thus, it is concluded that *L. stoechas* L. can be used as a memory enhancer. Further studies are needed to elaborate on its detailed mechanisms, in regards to enhancement of memory and toxicity profile.

Supplementary Materials: The following are available online at https://www.mdpi.com/article/10.3390/plants10061259/s1, Figure S1: Spectrum of analysis of AfL.s by GC-MS.

Author Contributions: Conceptualization, A.M. and R.A.; methodology, M.A.; software, U.F.G.; validation, R.A.M. and C.C.M.; formal analysis, A.M.; investigation, M.I.; resources, A.M.; data curation, E.B. and A.M.; writing—original draft preparation, A.M.; visualization, M.I. and E.B.; project administration. All authors have read and agreed to the published version of the manuscript.

Funding: This research received no external funding.

Institutional Review Board Statement: The study was conducted according to the guidelines of the Declaration of Helsinki, and approved by the Institutional Review Board (or Ethics Committee) of Government College University Lahore.

Informed Consent Statement: Not applicable.

Data Availability Statement: All of the data are available and can be produced upon request.

Conflicts of Interest: The authors declare no conflict of interest.

References

1. González-Tejero, M.; Casares-Porcel, M.; Sánchez-Rojas, C.; Ramiro-Gutiérrez, J.; Mesa, J.M.; Pieroni, A.; Giusti, M.; Censorii, E.; de Pasquale, C.; Della, A.; et al. Medicinal plants in the Mediterranean area: Synthesis of the results of the project Rubia. *J. Ethnopharmacol.* **2008**, *116*, 341–357. [CrossRef]
2. Benítez, G.; González-Tejero, R.; Mesa, J.M. Pharmaceutical ethnobotany in the western part of Granada province (southern Spain): Ethnopharmacological synthesis. *J. Ethnopharmacol.* **2010**, *129*, 87–105. [CrossRef]
3. Gilani, A.; Aziz, N.; Khan, M.; Shaheen, F.; Jabeen, Q.; Siddiqui, B.; Herzig, J. Ethnopharmacological evaluation of the anticonvulsant, sedative and antispasmodic activities of *Lavandula stoechas* L. *J. Ethnopharmacol.* **2000**, *71*, 161–167. [CrossRef]
4. Hakeem, M.; Siddiqui, M.; Khan, A. Antiepileptic activity of Ustu khudoos in secondary epilepsy—A case report. *Hamdard Med.* **1991**, *34*, 33–39.
5. Bouyahya, A.; Et-Touys, A.; Abrini, J.; Talbaoui, A.; Fellah, H.; Bakri, Y.; Dakka, N. Lavandula stoechas essential oil from Morocco as novel source of antileishmanial, antibacterial and antioxidant activities. *Biocatal. Agric. Biotechnol.* **2017**, *12*, 179–184. [CrossRef]
6. Angioni, A.; Barra, A.; Coroneo, V.; Dessi, S.; Cabras, P. Chemical composition, seasonal variability, and antifungal activity of *Lavandula stoechas* L. ssp. stoechas essential oils from stem/leaves and flowers. *J. Agric. Food Chem.* **2006**, *54*, 4364–4370. [CrossRef]
7. Re, L.; Barocci, S.; Sonnino, S.; Mencarelli, A.; Vivani, C.; Paolucci, G.; Mosca, E. Linalool modifies the nicotinic receptor–ion channel kinetics at the mouse neuromuscular junction. *Pharmacol. Res.* **2000**, *42*, 177–181. [CrossRef] [PubMed]
8. Cavanagh, H.; Wilkinson, J. Biological activities of lavender essential oil. *Phytother. Res.* **2002**, *16*, 301–308. [CrossRef] [PubMed]
9. Prashar, A.; Locke, I.C.; Evans, C.S. Cytotoxicity of lavender oil and its major components to human skin cells. *Cell Prolif.* **2004**, *37*, 221–229. [CrossRef]
10. Bouzouita, N.; Kachouri, F.; Hamdi, M.; Chaabouni, M.M.; Ben Aissa, R.; Zgoulli, S.; Thonart, P.; Carlier, A.; Marlier, M.; Lognay, G.C. Volatile Constituents and Antimicrobial Activity of *Lavandula stoechas* L. Oil from Tunisia. *J. Essent. Oil Res.* **2005**, *17*, 584–586. [CrossRef]
11. Zuzarte, M.; Gonçalves, M.; Cavaleiro, C.; Cruz, M.; Benzarti, A.; Marongiu, B.; Maxia, A.; Piras, A.; Salgueiro, L. Antifungal and anti-inflammatory potential of Lavandula stoechas and Thymus herba-barona essential oils. *Ind. Crop. Prod.* **2013**, *44*, 97–103. [CrossRef]
12. Benabdelkader, T.; Zitouni, A.; Guitton, Y.; Jullien, F.; Maitre, D.; Casabianca, H.; Legendre, L.; Kameli, A. Essential oils from wild populations of algerian *lavandula stoechas* l.: Composition, chemical variability, and in vitro biological properties. *Chem. Biodivers.* **2011**, *8*, 937–953. [CrossRef] [PubMed]
13. Goren, A.C.; Topçu, G.; Bilsel, G.; Bilsel, M.; Aydoğmusç, Z.; Pezzuto, J.M. The chemical constituents and biological activity of essential oil of *Lavandula stoechas* ssp. stoechas. *Z. Naturforsch. C* **2002**, *57*, 797–800. [CrossRef]
14. Sebai, H.; Selmi, S.; Rtibi, K.; Souli, A.; Gharbi, N.; Sakly, M. Lavender (*Lavandula stoechas* L.) essential oils attenuate hyperglycemia and protect against oxidative stress in alloxan-induced diabetic rats. *Lipids Health Dis.* **2013**, *12*, 189. [CrossRef] [PubMed]

15. Mushtaq, A.; Anwar, R.; Ahmad, M. *Lavandula stoechas* (L) a Very Potent Antioxidant Attenuates Dementia in Scopolamine Induced Memory Deficit Mice. *Front. Pharmacol.* **2018**, *9*, 1375. [CrossRef]
16. Davies, D.R.; Johnson, T.M. Isolation of three components from spearmint oil: An exercise in column and thin-layer chromatography. *J. Chem. Educ.* **2007**, *84*, 318. [CrossRef]
17. Ali-Shtayeh, M.S.; Jamous, R.M.; Abu Zaitoun, S.Y.; Qasem, I.B. In-vitro screening of acetylcholinesterase inhibitory activity of extracts from Palestinian indigenous flora in relation to the treatment of Alzheimer's disease. *Funct. Foods Health Dis.* **2014**, *4*, 381. [CrossRef]
18. Zhu, Y.; Zhang, S. Antibacterial activity and mechanism of lacidophilin from Lactobacillus pentosus against Staphylococcus aureus and Escherichia coli. *Front. Microbiol.* **2016**, *34*, 71–79.
19. Bobescu, E.; Covaciu, A.; Rus, H.; Radoi, M.; Badea, M.; Moga, S.N.; Benza, V.; Marceanu, L.G. Correlation of cardiovascular risk factors and biomarkers with platelet reactivity in coronary artery disease. *Am. J. Ther.* **2019**, *26*, E563–E569. [CrossRef]
20. Ulubelen, A.; Olcay, Y. Triterpenoids from *Lavandula stoechas* subsp. stoechas. *Fitoterapia* **1989**, *60*, 475–476.
21. Gohar, U.F.; Iqbal, I.; Shah, Z.; Mukhtar, H.; Zia-Ul-Haq, M. COVID-19: Recent Developments in Therapeutic Approaches. In *Alternative Medicine Interventions for COVID-19*; Zia-Ul-Haq, M., Bin-Jumah, M.N., Alothamn, S.I., Henidi, H.A., Eds.; Springer: Cham, Switzerland, 2021; pp. 249–274.
22. Orhan, I.; Aslan, M. Appraisal of scopolamine-induced antiamnesic effect in mice and in vitro antiacetylcholinesterase and antioxidant activities of some traditionally used Lamiaceae plants. *J. Ethnopharmacol.* **2009**, *122*, 327–332. [CrossRef] [PubMed]
23. Ademiluyi, A.O.; Ogunsuyi, O.B.; Oboh, G.; Agbebi, O.J. Jimson weed (*Datura stramonium* L.) alkaloid extracts modulate cholinesterase and monoamine oxidase activities in vitro: Possible modulatory effect on neuronal function. *Comp. Clin. Pathol.* **2016**, *25*, 733–741. [CrossRef]
24. Cometa, M.F.; Fortuna, S.; Palazzino, G.; Volpe, M.T.; Salgado, E.R.; Nicoletti, M.; Tomassini, L. New cholinesterase inhibiting bisbenzylisoquinoline alkaloids from Abuta grandifolia. *Fitoterapia* **2012**, *83*, 476–480. [CrossRef] [PubMed]
25. Arya, A.; Chahal, R.; Rao, R.; Rahman, M.H.; Kaushik, D.; Akhtar, M.F.; Saleem, A.; Khalifa, S.M.A.; El-Seedi, H.R.; Kamel, M.; et al. Acetylcholinesterase Inhibitory Potential of Various Sesquiterpene Analogues for Alzheimer's Disease Therapy. *Biomolecules* **2021**, *11*, 350. [CrossRef]
26. Lam, V.M.; Espinoza, S.; Gerasimov, A.S.; Gainetdinov, R.R.; Salahpour, A. In-Vivo pharmacology of trace-amine associated receptor 1. *Eur. J. Pharmacol.* **2015**, *763*, 136–142. [CrossRef] [PubMed]
27. Khan, M.Z.; Nawaz, W. The emerging roles of human trace amines and human trace amine-associated receptors (hTAARs) in central nervous system. *Biomed. Pharmacother.* **2016**, *83*, 439–449. [CrossRef]
28. Cichero, E.; Espinoza, S.; Gainetdinov, R.; Brasili, L.; Fossa, P. Insights into the Structure and Pharmacology of the Human Trace Amine-Associated Receptor 1 (hTAAR1): Homology Modelling and Docking Studies. *Chem. Biol. Drug Des.* **2013**, *81*, 509–516. [CrossRef]
29. Zia-Ul-Haq, M. Past, Present and Future of Carotenoids Research. In *Carotenoids: Structure and Function in the Human Body*; Zia-Ul-Haq, M., Dewanjee, S., Riaz, M., Eds.; Springer: Cham, Switzerland, 2021; pp. 827–854.
30. Wenk, G.L. Neuropathologic changes in Alzheimer's disease. *J. Clin. Psychiatry* **2003**, *64*, 7–10.
31. Da Silva Filho, S.R.B.; Barbosa, J.H.O.; Rondinoni, C.; dos Santos, A.C.; Salmon, C.E.G.; da Costa Lima, N.K.; Ferriolli, E.; Moriguti, J.C. Neuro-degeneration profile of Alzheimer's patients: A brain morphometry study. *NeuroImage Clin.* **2017**, *15*, 15–24. [CrossRef]
32. Omar, S.H.; Scott, C.J.; Hamlin, A.S.; Obied, H.K. The protective role of plant biophenols in mechanisms of Alzheimer's disease. *J. Nutr. Biochem.* **2017**, *47*, 1–20. [CrossRef] [PubMed]
33. McCord, J.M. The evolution of free radicals and oxidative stress. *Am. J. Med.* **2000**, *108*, 652–659. [CrossRef]
34. Praticò, D. Oxidative stress hypothesis in Alzheimer's disease: A reappraisal. *Trends Pharmacol. Sci.* **2008**, *29*, 609–615. [CrossRef] [PubMed]
35. Baxter, M.G.; Bucci, D.J.; Gorman, L.K.; Wiley, R.G.; Gallagher, M. Selective immunotoxic lesions of basal forebrain cholinergic cells: Effects on learning and memory in rats. *Behav. Neurosci.* **2013**, *127*, 619–627. [CrossRef]
36. Lu, C.; Wang, Y.; Wang, D.; Zhang, L.; Lv, J.; Jiang, N.; Fan, B.; Liu, X.; Wang, F. Neuroprotective effects of soy isoflavones on scopolamine-induced amnesia in mice. *Nutrients* **2018**, *10*, 853. [CrossRef] [PubMed]
37. Narang, D.; Tomlinson, S.; Holt, A.; Mousseau, D.D.; Baker, G.B. Trace amines and their relevance to psychiatry and neurology: A brief overview. *Klin. Psikofarmakol. Bül. Bull. Clin. Psychopharmacol.* **2011**, *21*, 73–79. [CrossRef]
38. Nelson, M.E.; Bryant, S.M.; Aks, S.E. Emerging Drugs of Abuse. *Emerg. Med. Clin. N. Am.* **2014**, *32*, 1–28. [CrossRef]
39. Ma, X.; Gang, D.R. In Vitro production of huperzine A, a promising drug candidate for Alzheimer's disease. *Phytochemistry* **2008**, *69*, 2022–2028. [CrossRef]
40. Dourish, C.T.; Cooper, S.J. Pharmacology of beta-phenylethylamine-induced seizures in mice. *Prog. Neuro Psychopharmacol. Biol. Psychiatry* **1983**, *7*, 787–790. [CrossRef]

plants

MDPI

Article

Aromatic Profile Variation of Essential Oil from Dried Makwhaen Fruit and Related Species

Trid Sriwichai [1,†], Jiratchaya Wisetkomolmat [1,2,†], Tonapha Pusadee [2,3], Korawan Sringarm [2,4], Kiattisak Duangmal [5,6], Shashanka K. Prasad [7], Bajaree Chuttong [2,8,*] and Sarana Rose Sommano [1,2,*]

1 Plant Bioactive Compound Laboratory (BAC Lab), Department of Plant and Soil Sciences, Faculty of Agriculture, Chiang Mai University, Chiang Mai 50200, Thailand; trid_s@cmu.ac.th (T.S.); jiratchaya_wis@cmu.ac.th (J.W.)
2 Innovative Agriculture Research Center, Faculty of Agriculture, Chiang Mai University, Chiang Mai 50200, Thailand; tonapha.p@cmu.ac.th (T.P.); kanok70@hotmail.com (K.S.)
3 Plant Genetic Resource and Nutrition Laboratory, Department of Plant and Soil Sciences, Faculty of Agriculture, Chiang Mai University, Chiang Mai 50200, Thailand
4 Department of Animal and Aquatic Sciences, Faculty of Agriculture, Chiang Mai University, Chiang Mai 50200, Thailand
5 Department of Food Technology, Faculty of Science, Chulalongkorn University, Bangkok 10330, Thailand; kiattisak.d@chula.ac.th
6 Emerging Process for Food Functionality Design Research Unit, Chulalongkorn University, Bangkok 10330, Thailand
7 Department of Biotechnology and Bioinformatics, Faculty of Life Sciences, JSS Academy of Higher Education and Research, Mysuru 570004, India; shashankaprasad@jssuni.edu.in
8 Meliponini and Apini Research Laboratory, Department of Entomology and Plant Pathology, Faculty of Agriculture, Chiang Mai University, Chiang Mai 50200, Thailand
* Correspondence: bajaree.c@cmu.ac.th (B.C.); sarana.s@cmu.ac.th (S.R.S.); Tel.: +66-53944040 (S.R.S.)
† These authors contributed equally as the first author.

Citation: Sriwichai, T.; Wisetkomolmat, J.; Pusadee, T.; Sringarm, K.; Duangmal, K.; Prasad, S.K.; Chuttong, B.; Sommano, S.R. Aromatic Profile Variation of Essential Oil from Dried Makwhaen Fruit and Related Species. *Plants* **2021**, *10*, 803. https://doi.org/10.3390/plants10040803

Academic Editor: Marc (Vlaic) Romina Alina

Received: 24 March 2021
Accepted: 16 April 2021
Published: 19 April 2021

Abstract: The aim of this research is to evaluate the relationship between genotype, phenotype, and chemical profiles of essential oil obtained from available *Zanthoxylum* spp. Three specimens of makhwaen (MK) distributed in Northern Thailand were genetically and morphologically compared with other *Zanthoxylum* spices, known locally as huajiao (HJ) and makwoung (MKO), respectively. HJ was taxonomically confirmed as *Z. armatum* while MKO and MK were identified as *Z. rhetsa* and *Z. myriacanthum*. Genetic sequencing distributed these species into three groups accordingly to their confirmed species. Essential oil of the dried fruits from these samples was extracted and analyzed for their chemical and physical properties. Cluster analysis of their volatile compositions separated MKO and MK apart from HJ with L-limonene, terpinen-4-ol, β-phellandrene, and β-philandrene. By using odor attributes, the essential oil of MKO and MK were closely related possessing fruity, woody, and citrus aromas, while the HJ was distinctive. Overall, the phenotypic characteristic can be used to elucidate the species among makhwaen fruits of different sources. The volatile profiling was nonetheless dependent on the genotypes but makwoung and makhwaen showed similar profiles.

Keywords: aromatic plant; chemical profiles; huajiao; spicy plant; taxonomical description; volatile compositions

1. Introduction

Plants of the *Zanthoxylum* spp. (Rutaceae) contain oil glands that yield high amounts of essential oil with distinctive aroma [1]. Their fruits are known as spices for ethnic food particular in Asia such as those of *Z. piperitum* [2], *Z. armatum* [3], *Z. fagara* [4], and the essential oils extracted from the fruits and leaves are used as food additives and functional ingredients in food and pharmaceutical industries. Commonly known as makhwaen or makhan, *Z. myriacanthum* is grown extensively in many areas of northern Thailand viz., Pong district of Payao, Song Khwae district of Nan and in many high-altitude areas of

Chiang Mai [5]. Previous studies described that *Z. myriacanthum* essential oil gives a unique citrus top-note followed by a woody and spice aromatic profile [4–6]. The analysis of the volatile compositions of *Z. myriacanthum* has illustrated that the main compounds are comprised of sabinene, terpinene-4-ol, and L-limonene [5,6]. Moreover, essential oils of plants in this genus also possess biological activities such as antimicrobial properties [7] antioxidant activity [8] and anti-inflammatory [9] thereby are used for medicinal purposes. With an increasing commercial need for exotic ingredients, there is, therefore demands for high-quality raw material of essential oil for food and perfumery industries. To the perfumery industry claim, the complaint made from the raw material purchaser asserted that plant morphological characteristics such as tree structure, sizes, and color of the berry clusters were variable in different sourcing regions which made the final quality of essential oil unsteady (Mrs. Anne Saget pers. Comm.). Moreover, the complexity within the species remains ambiguous as such *Z. myriacanthum* is often misidentified as *Z. limonella* [10]. Thus, there is urge commercially to truly describe plant species.

Genetic and environmental variables—i.e., growing condition, light intensity, day length, temperature, altitude, as well as their interactions—could generally influence the quantity and quality of the essential oils [11,12]. Identification of plant species and variety in the same genus can be accomplished by taxonomic description and chemical compositions [13]. However, only the use of these phenomena may not be enough to accurately describe the species. Studies on the essential oil containing plants revealed that chemical compositions and characteristics of essential oils from plants within the same genus are diverse such as those belonging to *Ocimum* spp. [14] and *Zanthoxylum* spp. [15]. The use of DNA fingerprints can therefore accomplish for the reliable identification of plant species [16].

Internal transcribed spacer DNA barcode (ITS2) detects nuclear marker of the rDNA region in nuclear genome that is useful for directly detecting reticulate phenomena. This technique has been reported to be an efficient barcode locus for plant identification [17,18] and classification by many plant species such as Indian *Berberis* [19], timber species of the mahogany family [20], and *Dendrobium* species [21]. In addition to the ITS region, RAPD analysis is an alternate method for estimating genetic diversity and relatedness in plant populations, cultivars and germplasm accessions, especially in non-model plant species. By using the markers, the technique is able to amplify DNA from dispersed polymorphic loci and thereby indirectly distinguishes small differences within the gene sequences. To draw accurate conclusion on genetic relations of plants species, it is therefore vital to combine these techniques. There is no research work to-date that fully describe genotyping differences among raw materials for makhwaen essential oil production in relation to their physical properties and aromatic profiles as compared to those of other *Zanthoxylum* species. The aim of this research, therefore, is to descriptively establish profile specification of raw materials used in makhwaen essential oil extraction industry.

2. Results and Discussion

2.1. Morphological Confirmation

The morphological descriptions of the specimens of the *Zanthoxylum* spp. known locally as huajiao (HJ), makwoung (MKO), and makwhaen (MK1-3) were documented using plant structure, thorn, leaf type, floral structures, and fresh fruit color [22]. From our data in Supplementary Table S1, the plant structure of HJ was of shrub and was different from that of MKO and MK1-3 (tree-like structure). Thorns of all specimens were either initiate on the trunk or branches. The same compound leaf type was observed in the MKO and MK1-3 (even-pinnately) which were different from the HJ (odd-pinnately compound leaf). The floral compositions were different in every species; the HJ consisted of the flower with six to nine petals, while the MKO was with four petals and MK1-3 were with five petals. Within a similar pattern, the number of anthers was different in every species, four to eight anthers for HJ, three or four anthers for MKO, and five anthers for MK1-3. The color of fresh fruit was red in HJ and MKO while MK1-3 gave greenish-red color characteristics. Fruit sizes varied from 2–3 mm of the MK1-3 to 4–5 mm of the HJ and the MKO was 5–7 mm, respectively. The three species gave brown fruit when dried with crack revealing the inner seeds. According to these specific characteristics, the scientific names of *Z. armatum*, *Z. rhetsa*, and *Z. myriacanthum* are given to HJ, MKO, and MK specimens [22,23]. To describe the verity within the same species, floral and fruit characteristics of MK1-3 were compared (Supplementary Table S2). The result confirmed that the MKs were those of *Z. myriacanthum* as the sepals and petals were pentamerous and male flower organs composed five stamens.

According to the results from UPGMA analysis of plant characteristics and seven samples of *Zanthoxylum* spp., MK1-3 were far distinctive from MKO and HJ. As it could be seen, MKO and MK1-3 were both tree plants while HJ was a shrub. Nonetheless, MK1-3 were detached from MKO by their floral characteristics (Figure S1).

2.2. ITS Sequencing Analysis

The aligned lengths of the ITS region (including both ITS1 and ITS2 regions) ranged from 596 bp for MK (*Z. myriacanthum*) to 600 bp for HJ (*Z. armatum*). Among the five MK sampling (two specimens from Mae Tang district: MK1-1 and MK1-2, one from Mae Rim: MK2 and two from Nan: MK3-1 and MK3-2), the ITS sequences were completely identical whereas 39 single nucleotide polymorphism were found among MK, MKO, and HJ samples. The phylogenetic relationship analysis was investigated based on the total ITS region sequences. The dendrogram showed three major clades (Figure 1A), the first formed among five MK samples from the three regions, the second consisted of MKO while the last is HJ. ITS sequence is an efficient tool for genetic identification among species however, very low efficiency for evaluation of genetic variation within the species.

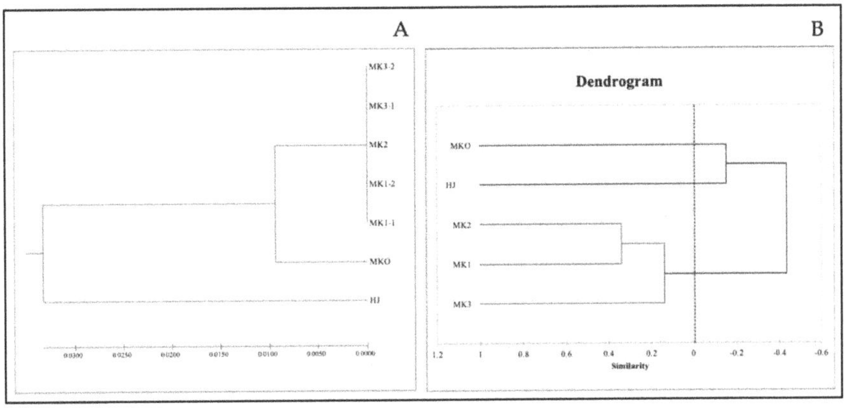

Figure 1. The dendrogram of *Zanthoxylum* spp. in North of Thailand; HJ (huajiao), MKO (makwoung), MK1 (makhwaen from Mae Tang district), MK2 (makhwaen from Mae Rim district), and MK3 (makhwaen from Song Kwae district) derived by UPGMA from the similarity matrix of the ITS sequence data (**A**) and from the similarity matrix based on 37 DNA bands obtained from five RAPD markers (**B**).

2.3. RAPD Analysis

Only S6, S7, S9, OPA01, and OPA04 primers gave responses with the DNA thus they were used for the calculation of the unweighted pair group method with arithmetic mean (UPGMA). Result was illustrated as a dendrogram in which the samples were split into three groups: group 1—HJ, group 2—MKO, and group 3 consisting of MK1-3, as shown in Figure 1B. The dendrogram illustrated that MK1-3 were clustered as closely related species while HJ and MKO were genetically identified as separated species. Indeed, the RAPD makers revealed slight genetic variables between the *Z. myriacanthum* samples from different geographical regions. The result was correspondent with our taxonomic data described previously. In addition to the ITS sequencing, the RAPD technique was successful to determine the genetic variation of the Zanthoxylum spp. as well as many plants of this kind including *Z. hamiltonianum, Z. nitidum, Z. oxyphyllum, Z. rhesta, Z. armatum*, and *Z. schinifolium* [24–26].

2.4. Essential Oil Analysis

Essential oils were extracted from dried fruits of makhwaen samples from three areas (MK1, MK2, and MK3), huajiao (HJ) and makwoung (MKO) using hydro-distillation. The extraction yield varied by mean of species differentiation i.e., MK1-3 (~7%), followed by HJ (~5%) and MKO (~2%). Thirty-five volatile compounds were detected using GC-MS (Table 1). Essential oil of the MKO contained the major content of linalool (7.35 $\mu g \cdot mL^{-1}$) following by β-thujone (1.03 $\mu g \cdot mL^{-1}$) and sabinene (0.44 $\mu g \cdot mL^{-1}$), respectively. Sabinene was the key dominant substance in the *Zanthoxylum* species analyzed, except for the essential oil of the MKO. This is in agreement with other works done with plants belongs to the *Zanthoxylum* species—i.e., *Z. xanthoxyloïdes, Z. leprieurii* [27], and *Z. rhoifolium* [28] with sabinene and limonene that represented woody and citrus aromas [29].

The chemical profiles of the essential oils from makhwaen fruits collected from different locations were variable. The major components of all samples could be described as following sequence: MK1; limonene (4.05 $\mu g \cdot mL^{-1}$), sabinene (3.20 $\mu g \cdot mL^{-1}$) and L^{-}phellandrene (1.47 $\mu g \cdot mL^{-1}$), MK2; sabinene (2.55 $\mu g \cdot mL^{-1}$), terpinen-4-ol (2.05 $\mu g \cdot mL^{-1}$) and β-phellandrene (1.85 $\mu g \cdot mL^{-1}$), MK3; limonene (6.89 $\mu g \cdot mL^{-1}$), sabinene (3.00 $\mu g \cdot mL^{-1}$) and β-ocimen (1.47 $\mu g \cdot mL^{-1}$), HJ; sabinene (4.56 $\mu g \cdot mL^{-1}$), terpinen-4-ol (4.31 $\mu g \cdot mL^{-1}$) and γ-terpene (1.08 $\mu g \cdot mL^{-1}$). To this extend, geographical or environmental factors would play an important role in the chemical composition of the volatiles [12]. The variations due to growing locations of aromatic crops were fully described in chamomile (*Matricaria recutita* L.) [30], *Satureja kitaibelii* [31] and *Myrsine leuconeura* [32]. In the *Zanthoxylum* spp., plants growing at different altitudes yielded essential oil with alternating principal volatiles (limonene, sabinene, and linalool) viz., *Z. armatum* [3,33,34] and *Z. alatum* [35]. Our results agree with this as plant samples taken for this experiment were grown at different altitudes.

The relationships between the chemical components and the *Zanthoxylum* species were analysed using the PCA in Figure 2. The PCA revealed that HJ was distinctive from the other *Zanthoxylum* spp. and MKO could not be detached from MK (PC1 40.78% and PC2 20.49%). According to the bi-plot (Figure 2b), L-linalool was principal in the HJ while L-limonene, terpinen-4-ol, and β-phellandrene were among the major components found in MK and MKO. By interpreting the volatile substances according to their descriptors using a heatmap, it was found that HJ was also separated from other species (Figure 3) with different aromatic profile patterns.

Demands of high-quality essential oil from raw material of unique plant taxa for food and perfumery production has ramped up recently. Essential oil compositions could assist in genetic analysis of plant species thus the generic term of chemotypes is well perceived [36,37]. Based on our result of the chemometric analyses, L-linalool was separated from the others and projected with the HJ similar with the result form the RAPD analysis. Therefore, it could be used indirectly as a marker for characterization of the *Zanthoxylum* species. More importantly, the heat mapping of the odor descriptors also convinced that of all the analyzed species, HJ represents the citrusy-floral aroma which is its unique aroma identity. This has been described as the generic perception of Sichuan pepper aroma [38]. Besides, the volatile compositions, non-volatiles such as alkylamides and polyphenols are known as specific chemotypes of the *Zanthoxylum* spp. These compounds offer spice flavor with tingling and numbing sensations [37–39].

Table 1. Chemical profiles of makhwaen, huajiao, and makwoung essential oils

No.	Chemical Compounds	Descriptors [5]	RI^{cal}	RI^{ref}	Amount of Chemical (μg.mL^{-1} Essential Oil #)				
					MK1	MK2	MK3	HJ	MKO
The Amount of Essential Oil Extractions (%)					7.15 ± 0.07	6.2 ± 0.14	7.4 ± 0.14	5.5 ± 0.01	1.8 ± 0.07
1	α-Thujene	Woody	926	926	0.15 ± 0.01	0.11 ± 0.02	0.04 ± 0.01	0.18 ± 0.02	0.05 ± 0.02
2	α-Pinene	Pine	937	1004	0.42 ± 0.02	0.45 ± 0.02	0.09 ± 0.01	0.52 ± 0.02	0.03 ± 0.02
3	Sabinene #	Woody	942	937	3.20 ± 0.18	2.55 ± 0.02	3.00 ± 0.08	4.56 ± 0.02	0.44 ± 0.02
4	2β-Pinene	Pine	974	944	ND	0.02 ± 0.02	ND	0.75 ± 0.02	0.05 ± 0.02
5	β-Myrcene	Spicy	993	1132	0.66 ± 0.04	0.27 ± 0.02	0.86 ± 0.05	0.21 ± 0.02	0.13 ± 0.02
6	Octanal	Citrus	999	992	0.06 ± 0.01	0.05 ± 0.02	0.10 ± 0.01	0.02 ± 0.02	ND
7	L-Phellandrene	Fruity	1009	989	1.47 ± 0.07	0.61 ± 0.02	0.32 ± 0.02	0.09 ± 0.02	0.04 ± 0.02
8	Acetic acid, Hexyl ester	Sweet/Floral	1048	1131	0.02 ± 0.01	0.04 ± 0.02	0.02 ± 0.01	ND	ND
9	α-Terpinene	Citrus	1018	1196	0.29 ± 0.02	0.40 ± 0.02	0.08 ± 0.01	ND	ND
10	Methyl (1-methylethyl)-benzene	Citrus	1058	1058	0.16 ± 0.01	0.91 ± 0.02	0.03 ± 0.01	0.50 ± 0.02	0.13 ± 0.02
11	L-Limonene #	Citrus	1047	1035	4.05 ± 0.01	1.01 ± 0.02	6.89 ± 0.18	0.31 ± 0.02	0.97 ± 0.02
12	β-Phellandrene	Fruity	1103	1227	1.08 ± 0.08	1.85 ± 0.02	0.90 ± 0.05	0.42 ± 0.02	ND
13	*cis*-Ocimene	Herbal	1132	1132	0.17 ± 0.01	0.11 ± 0.02	0.08 ± 0.01	0.06 ± 0.02	0.02 ± 0.02
14	β-Ocimene	Herbal	1144	1017	0.76 ± 0.04	0.31 ± 0.02	1.47 ± 0.06	ND	0.03 ± 0.02
15	γ-Terpinene	Fruity	1168	1168	0.47 ± 0.02	0.63 ± 0.02	0.14 ± 0.01	1.08 ± 0.02	0.25 ± 0.02
16	*trans* Sabinene hydrate	Herbal	1180	1458	0.06 ± 0.01	ND	0.06 ± 0.01	0.39 ± 0.02	ND
17	L-Octanol	Waxy	1167	1578	0.09 ± 0.01	0.17 ± 0.02	0.05 ± 0.01	0.03 ± 0.02	ND
18	α-Terpinolene	Fruity	1236	1236	0.15 ± 0.01	0.17 ± 0.02	0.09 ± 0.01	0.30 ± 0.02	ND
19	L-Linalool	Sweet/Floral	1263	1594	0.43±0.02	0.53 ± 0.02	0.26 ± 0.02	ND	7.35 ± 0.02
20	L-Terpineol	Woody	1189	1387	0.07 ± 0.01	0.11 ± 0.02	0.09 ± 0.07	0.21 ± 0.02	0.08 ± 0.02
21	Terpinen-4-ol	Citrus	1391	1566	1.09 ± 0.07	2.05 ± 0.02	0.39 ± 0.03	4.31 ± 0.02	0.14 ± 0.02
22	Sabina ketone	Minty	1236	1161	0.03 ± 0.01	0.06 ± 0.02	ND	ND	ND
23	β-Thujone	Minty	1251	1011	ND	ND	ND	ND	1.03 ± 0.02
24	β-Fenchyl alcohol	Pine	1263	1130	0.37 ± 0.08	0.53 ± 0.02	0.13 ± 0.08	0.27 ± 0.02	0.16 ± 0.02
25	*trans*-Piperitol	Herbal	1282	1210	0.02 ± 0.02	ND	ND	0.12 ± 0.02	ND
26	Decanal	Sweet/Floral	1423	1423	0.22 ± 0.02	0.16 ± 0.02	0.24 ± 0.01	0.17 ± 0.02	ND
27	Acetic acid, 2-Ethylhexyl ester	Herbal	1370	-	0.28 ± 0.01	0.48 ± 0.02	0.25 ± 0.02	ND	ND
28	*trans*-Geraniol	Sweet/Floral	1442	1442	0.03 ± 0.01	0.12 ± 0.02	ND	ND	0.55 ± 0.02
29	1-Decanol	Sweet/Floral	1457	1457	0.03 ± 0.03	0.07 ± 0.02	0.04 ± 0.01	0.06 ± 0.02	ND
30	2-Undecanone	Fruity	1467	1467	0.47 ± 0.02	0.04 ± 0.02	0.36 ± 0.04	ND	ND
31	Geranyl acetate	Sweet/Floral	1408	1408	0.26 ± 0.01	0.23 ± 0.02	0.03 ± 0.01	ND	ND
32	Dodecanal	Citrus	1439	1439	0.07 ± 0.01	0.05 ± 0.02	0.06 ± 0.01	ND	ND
33	*trans*-Caryophyllene	Spicy	1441	1441	0.03 ± 0.03	0.02 ± 0.02	0.03 ± 0.01	0.09 ± 0.02	ND
34	D-Germacrene	Woody	1447	1447	0.09 ± 0.01	ND	ND	0.03 ± 0.02	ND
35	Bicyclogermacrene	Woody	1457	1457	0.02 ± 0.02	ND	ND	ND	ND
	Total			-	16.76	13.09	16.09	14.68	11.44

RI^{cal}: Calculated retention index. RI^{Ref}: Retention index from the referent [5]. # Values are calculated as a reference to internal standard toluene (0.003% $w{\cdot}v^{-1}$). Makhwaen fruit, huajiao and makwhoung essential oil were analyzed by GC-MS (MK1, MK2, MK3, HJ, and MKO). ND: not detected.

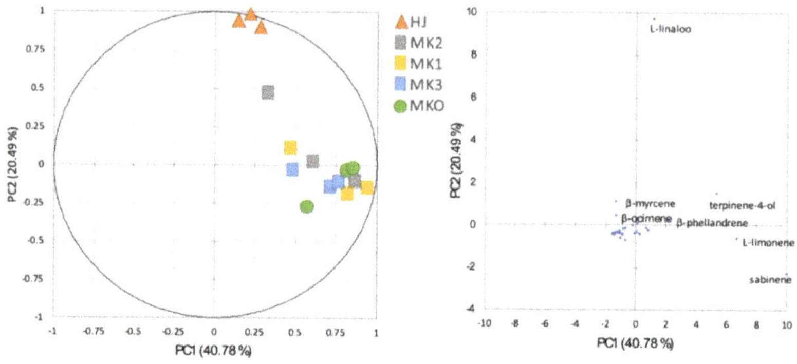

(**a**) (**b**)

Figure 2. Principal component analysis (PCA) illustrating the relationships among the *Zanthoxylum* species (**a**) and bi-plot factor analysis of the chemical components of the *Zanthoxylum* essential oils (**b**). Abbreviations; HJ (huajiao), MKO (makwoung), MK1 (makhwaen from Mae Tang district), MK2 (makhwaen from Mae Rim district), and MK3 (makhwaen from Song Kwae district).

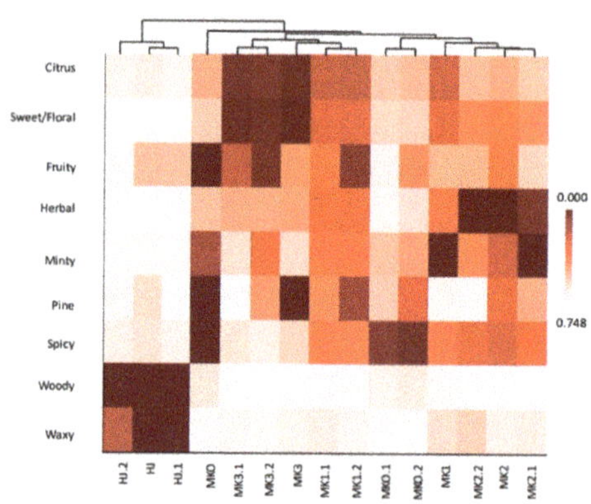

Figure 3. Heatmap relationship of the odor descriptors representing the volatile composition of the *Zanthoxylum* essential oils. Abbreviations; HJ (huajiao), MKO (makwoung), MK1 (makhwaen from Mae Tang district), MK2 (makhwaen from Mae Rim district), and MK3 (makhwaen from Song Kwae district).

2.5. FTIR Analysis

Fourier transform infrared spectroscopy (FTIR) spectrum patterns have been adopted to expose authentically volatile composition of plant essential oils such as those of lavender (*Lavandula officinalis*), pepper-mint (*Mentha piperita*), green doulas (*Pseudotsuga menziesii*), fir (*Abies alba*), and chicory (*Cichorium intybus*) [40,41]. The spectrum patterns of their EOs responded to the wavenumber ranges 2800–2300 and 1800–1000 cm^{-1}) representing of free O-H bond valence and carboxylic acid broadband absorption. Our results illustrated that the oil samples were dominated by overtones and different combinations of C-H reflection and shine occurring between 500–4000 cm^{-1} and aromatic ring at ~1600 cm^{-1}. FTIR spectrum scans of the three *Zanthoxylum* species essential oil (MK1-3, HJ, and MKO) absorbed light at a wavenumber range of 1722–798 cm^{-1} and 2967–2926 cm^{-1}, respectively, therefore illustrating similar light transmission. EO of the HJ on the other hand showed distinct spectrum characteristics from other samples (Figure 4 and Table 2). This distinction was in parallel with the odor descriptions above analyzed where HJ was indicated to have a sweet and floral scent.

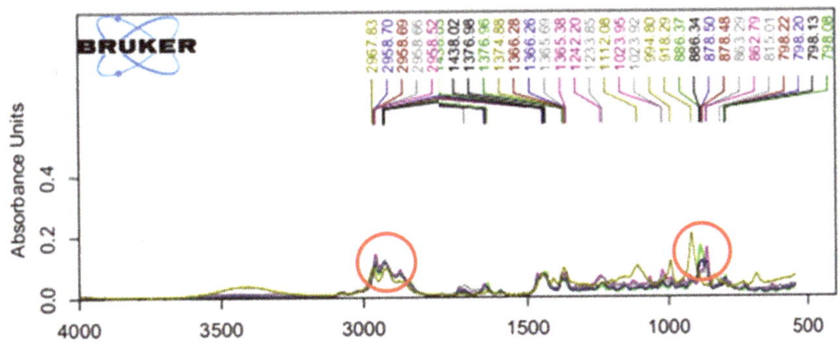

Figure 4. Fourier transform infrared spectrophotometer (FTIR) spectra of the essential oils from five different *Zanthoxylum* species. The insertion is the inset evidence of the peaks between 500–4000 cm^{-1}: (—) MK1, (—) MK2, (—) MK3, (—) HJ, and (—) MKO. Abbreviations; huajiao (HJ), makwoung (MKO), MK1 (makhwaen from Mae Tang district), MK2 (makhwaen from Mae Rim district), and MK3 (makhwaen from Song Kwae district).

Table 2. Wavenumbers and functional groups of *Zanthoxylum* spp. essential oils.

Name	Wavenumber (cm^{-1})	Type of Vibration	Functional Groups
HJ	918.29	(=C-H) bending strong	Alkene
	994.8	(=C-H) bending strong	Alkene
	1112.08	C-O stretch strong	Alcohol
	1374.88	bending variable -C-H	Alkane
	2967.83	C-H stretch strong	CH_2 group
MKO	862.79	(=C-H) bending strong	Alkene
	1023.95	C-O stretch strong	Alcohol
	1365.38	bending variable -C-H	Alkane
	1445.39	bending variable -C-H	Alkane
	2958.52	C-H stretch strong	CH_2 group
MK1	878.5	(=C-H) bending strong	Alkene
	1366.26	bending variable -C-H	Alkane
	1445.3	bending variable -C-H	Alkane
	1601.0	-	Aromatic ring
	1650.64	(C=O) stretch	Ester
	2958.7	C-H stretch strong	CH_2 group
MK2	863.29	(=C-H) bending strong	Alkene
	1233.85	(C-O) stretch	Alcohol
	1365.69	bending variable -C-H	Alkane
	1446.33	bending variable -C-H	Alkane
	2958.66	C-H stretch strong	CH_2 group
MK3	886.37	(=C-H) bending strong	Alkene
	1376.96	bending variable -C-H	Alkane
	1438.03	bending variable -C-H	Alkane
	1564.0	-	Aromatic ring
	1643.61	(C=O) stretch	Alcohol
	2926.69	C-H stretch strong	CH_2 group

Abbreviations: huajiao (HJ), makwoung (MKO), and makhwaen (MK1-3).

3. Materials and Methods

3.1. Plant Materials

Three plant specimens of the *Zanthozylum* spp. locally known as makhwaen were collected from the local orchards in three areas: (MK1) Papea, Mae Tang district, Chiang Mai province (19°7′27″ N, 98°42′14″ E); (MK2) Pong Yang, Mae Rim district, Chiang Mai province (18°53′24″ N, 98°49′53″ E); (MK3) Yod, Song Kwae district, Nan province (19°22′37″ N, 100°35′49″ E) in September 2018. Huajiao (HJ) was harvested from Ban Rak Thai, Mok Champae, Muang district, Mae Hong Sorn province (19°32′32″ N, 97°53′35″ E) in September 2018. Makwoung specimen (MKO) was sampled from Phichai, Muang Lampang District, Lampang province (18°22′11″ N, 99°35′44″ E) in September 2018 (Table S3). Based on the samples from harvest, all samples can be divided into two groups: (i) young leaves for DNA analysis and (ii) fruits for essential oil analysis.

The morphological appearances of leaves, flower, and fruit were recorded [19,41]. Their fruits correspondent to all specimen samples were also collected for the essential quality assessment at the mature stage and subjected to the initial drying process as described in the previous report [5]. A taxonomical confirmation has been done by comparison of the taxonomical descriptions with those of the literature data [22] and also confirmed by a botanist. The sample specimens were deposited at Queen Sirikit Botanic Garden (QSBG, Mae Rim, Chiang Mai, Thailand) and the accession numbers of Trid01-05C were assigned.

3.2. Morphology Relationship within Species of *Zanthoxylum* spp.

Collected data of the part of Plant for classification were analyzed. Those characters were assigned and scored as plant structure: shrub = 0, tree = 1; thorn: not have thorn = 0, thorn on tree = 1; compound leaf type: odd-pinnate = 0, even-pinnate = 1; number of petals: four petals = 0, five petals = 1, more than six petals = 2; number of anthers: four anthers = 0, five anthers = 1, more than six anthers = 2; fresh fruit color: red = 0, greenish-red = 1, and dry fruit color: brown = 0, no brown = 1. These data were analyzed using cluster analysis (Dendrogram and PCA-biplot) via XLstat, version 2016.

3.3. ITS and RAPD Analysis

3.3.1. DNA Extraction

For the extraction of DNA, the DT-S DNA extraction kit (Kurabo, Osaka, Japan) was used with modification of the CTAB extraction procedure. Young leaf tissue of three *Zanthozylum* spp. from five samples (0.5 g) were ground to powder using a mortar and pestle in the presence of liquid nitrogen and transferred to a 1.5 mL polypropylene centrifuge tubes and follow the steps of the DNA extraction kit. Tissue lysis-buffer (MDT) 200 μL and proteinase K (EDT) 20 μL were combined and mixed. After that, the centrifuge tubes were incubated by using the incubator at a temperature of 55 °C for an hour. At this stage, the centrifuge tubes were flipped every 15 min. Then, these tubes were centrifuged at 10,000× *g*. When the process was completed, the supernatant (~200 μL) was moved to a new centrifuge tubes and 180 μL lysis buffer (LDT) was added. Later, these new tubes were centrifuged with vortex for 15 s. before they were incubated at a temperature of 70 °C for 10 min. A solution was moved into the new cartridge tubes and west tubes, then these tubes were aerated. After that, 75 μL wash buffer (WDT) was added into the tubes. These tubes were aerated repeatedly three times to elute DNA. Then, the cartridge tubes were moved into the collection tubes. At this stage, 50 μL elution buffer (CDT) was added and left for 30 min. After that, they were aerated repeatedly for two times. Finally, the centrifuge tubes were tested and stored at a temperature of −20 °C.

After extraction, total DNA was quantified using a nano-drop spectrophotometer (NanoDrop^{TM} 1000 Spectrophotometer, Thermo Fisher Scientific, Bath, UK). For re-quantification, the extracted DNA was run on 1.5% agarose gel electrophoresis using 1× TBE buffer at 5–8 V·mL^{-1} for 30 min and visualized under BLook LED transilluminator (Genedirex, Taoyuan, Taiwan) by staining with MaestroSafe TM (Maestrogen,

Las Vegas, NV, USA). The DNA solution was diluted with sterile distilled water (DI) to a concentration of 10 ng·μL^{-1} for PCR analysis and kept at −20 °C until use [42].

3.3.2. ITS Sequence

The ITS2 sequences were amplified using the following pair of universal primers, ITS5-ITS4 (including both ITS1 and ITS2 regions), ITS5 GGAAGTAAAAGTCGTAACAAGG and ITS4 TCCTCCGCTTATTGATATGC. Each 50 μL reaction contained 5 μL 10× PCR buffer, 2.5 μL 2.5 mm $MgCl_2$, 0.4 μL 0.2 mm deoxyribonucleotides (dNTP), 5 μL of each primer (10 ng·μL^{-1}), 0.4 μL 0.5 U Taq DNA polymerase (HIMEDIA, Mumbai, India), 40 μL sterile distilled water, and 5 μL genomic DNA (50 ng·μL^{-1}). The amplification consisted of 94 °C·2 min^{-1}, followed by 40 cycles of 94 °C·45 s^{-1}, 50 °C·45 s^{-1}, and 72 °C·1 min^{-1}, and ending with 72 °C for 5 min for the final extension. Amplified products were genotyped using 1.5% agarose gel electrophoresis. Then they were staining with MaestroSafeTM Nucleic Acid Stains (MAESTROGEN, Hsinchu, Taiwan) and visualized under UV transilluminator (BioDoc-It2 imaging systems, Analytik Jena, Thuringia, Germany) before samples were sent to sequencing at Macrogen, Inc. (Seoul, South Korea).

3.3.3. RAPD-PCR Protocols

For RAPD analysis of the genomic DNA, 10-base primers from Operon Technologies (Alameda, GA, USA) and UBC (University of British Columbia, Canada) were chosen (Table S4). A total of nine primers from previous studies were screened [43–46]. The polymerase chain reaction (PCR) was adjusted to 10 μL^{-1} containing 8 μL^{-1} of OnePCRTM Plus (Genedirex, Taoyuan, Taiwan), 1 μL^{-1} of 1 μm RAPD primer and 1 μL^{-1} of 10 ng genomic DNA. All the reactions were carried out on a Flexcycler2 thermal cycler (Analytik Jena, Thuringia, Germany) using the following profile: 1 cycle, 94 °C, 4 min; 40 cycles, 95 °C, 30 s; 37 °C, 30 s; 72 °C, 60 s; 1 cycle, 72 °C, 10 min. The sample was separated in a 1.5% agarose gel in 1× TBE buffer. The samples were run at 70 V for 120 min. The gels were then visualized using the BLooK LED transilluminator (Genedirex, Taoyuan, Taiwan).

3.4. Dendrogram Analysis

The banding pattern for each primer was scored as diallelic (1 = band present, 0 = band absent), and stored in an Excel (Microsoft) spreadsheet file in the form of a binary matrix. To determine the genetic differentiation between the five samples accessions, 10 RAPD markers were analyzed using the statistical package XLSTAT version 2016 software. The coefficients of genetic similarity for all the pair-wise comparisons were computed using Jaccard's coefficient of similarity and then the distance matrix was subjected to cluster analysis by using the Unweighted Pair Group Method with Arithmetic Mean (UPGMA) to produce a dendrogram.

3.5. Essential Oil Analysis

The essential oil was extracted by hydro-distillation for 4 h, from 100 g of dried fruits in 600 mL of DI water in a 2 L flask Clevenger-type apparatus (MTopo®, heating mantle, Korea). The oil was dried over anhydrous sodium sulphate (Merck Co., Darmstadt, Germany) and was kept at 4 °C until analysis (usually within three days). The extraction was repeated twice and yield (mean value) was reported as a percentage of essential oil from dry plant material [33].

Gas chromatography-mass spectrometry (GC–MS) analysis was performed on Bruker-Scion 436 GC (Bruker, Hamburg, Germany) a Rxi 5Sil MS (30 m × 0.25 mm; 0.25 μm film thickness) (Restek, Bellefonte, PA, USA). Essential oil samples (2 μL at the dilution of 1%, $v{\cdot}v^{-1}$, in dichloromethane (RCI Labscan, Bangkok, Thailand) with a presence of 0.003% $w{\cdot}v^{-1}$ toluene (RCI Labscan, Bangkok, Thailand) as an internal standard) were injected in a split mode (1:20). Temperature program includes oven temperature held for 2 min at 60 °C and was enhanced to 150 °C with 3 °C min^{-1}. Then, temperature enhancement was programmed up to 270 °C at 5 °C min^{-1} and held at this temperature for 15 min. Other operating conditions include carrier gas was Helium with a flow rate of 1.1 mL min^{-1}; injector and detector temperatures were 300 °C, and split ratio, 1:50. Mass spectra (MS), 50–500 ($m \cdot z^{-1}$) were taken at 70 eV. The mass spectra and retention indices of essential oil components were identified by comparison to MS computer library (NIST 05.L and NIST 98.L. Homologous series of C_8–C_{20} n–alkanes (Sigma–Aldrich, Steineheim, Germany) were used for identification of all constituents by calculation of the retention indexes (RI). The compounds were confirmed by their RI as well as those from the literature [14]. The amount in $\mu g \cdot mL^{-1}$ of essential oil was calculated as relative to that of internal standard.

3.6. *Fourier Transforms Infrared Spectrophotometer (FTIR) Analysis*

The FTIR spectrometer used was Bruker model ALPHA II, Diamond ATR (Hamburg, Germany) and operating at the basic of 500–4000 cm^{-1} wavenumbers for averaging 47 scans per spectrum [40].

3.7. *Statistical Analysis*

The data were statistically analyzed using a comparison of the means of yield for essential oils evaluated by Tukey Multiple Comparison's test at a 95% confidential level. A principal component analysis (PCA) was used to identify the main sources of systematic variation in the chemical compounds data using XLstat software version 2016 [5]. The amount of each volatiles was combined according to their descriptors as described in Sriwichai et al. [6] which then was used to explicate the odor profile of the essential oil. Heatmap was generated with Biovinci software (BioTuring Inc., San Diego, CA, USA).

4. Conclusions

Even though a large number of secondary metabolites interfere with DNA sequencing, morphological description is adequate for the differentiation of plant belonging to the *Zanthoxylum* genus. The locally known makhwaen were taxonomically and genetically confirmed as *Z. myriacanthum*. From the principal component evaluation, huajiao essential oil was described to have different aroma characteristic as compared to the rest of *Zanthoxylum* spp. analyzed. The essential oils of makwoung and makhwaens from Nan and Chiang Mai were similar in terms of quantity and characteristics of the chemical compositions. For example, limonene and sabinene represent the aroma of citrus and woody. In summary, for sourcing of the raw material, phenotypical characteristic can be used to distinguish the species. Furthermore, the chemical profile of the essential oil depends upon the genotypes which closer similarity was with makwoung and makhwaen, whereas huajiao represented the unique chemotype of citrus-floral aroma.

Supplementary Materials: The following are available online at https://www.mdpi.com/article/10.3390/plants10040803/s1, Figure S1: The dendrogram of *Zanthoxylum* spp. in North of Thailand; huajiao (HJ), makwhoung (MKO), MK1 (makhwaen from Mae Tang district), MK2 (makhwaen from Mae Rim district) and MK3 (makhwaen from Song Kwae district) derived by UPGMA from the similarity matrix based on seven morphology data (plant structure, thorn, compound leaf type, petals, anthers, fresh and dry fruit color); Table S1: Plant characteristics for taxonomical identification of collected *Zanthoxylum* spps. used in this experiment; Table S2: Floral and fruit characteristics of makhwaen collected from different locations (MK1-3); Table S3: Study site the sample collections; Table S4: Sequence of RAPD primers.

Author Contributions: Conceptualization, T.P. and S.R.S.; Methodology, T.P. and S.R.S.; Validation, T.S. and J.W.; Formal analysis, T.S. and J.W.; Investigation, K.D.; Data curation, T.S. and J.W.; Writing—original draft preparation, T.S., J.W., and S.R.S.; Writing—review and editing, S.R.S., K.D., S.K.P., and B.C.; Visualization, T.S. and J.W.; Supervision, T.P. and S.R.S.; Funding acquisition, K.S., B.C. and K.S. All authors have read and agreed to the published version of the manuscript.

Funding: This research work was partially supported by Chiang Mai University.

Institutional Review Board Statement: Not applicable.

Informed Consent Statement: Not applicable.

Data Availability Statement: Not applicable.

Acknowledgments: We would like to acknowledge the external supports from Cosmo Ingredients, Mougins, France. We also appreciate the generosity of Anne Marie Saget and Wei Raksa during course of experimentation.

Conflicts of Interest: The authors declare no conflict of interest.

References

1. Zhang, D.; Hartley, T.G.; Mabberley, D.J. Rutaceae. In *Flora of China 11 (Oxalidaceae through Aceraceae)*; Wu, Z.Y., Raven, P.H., Hong, D., Eds.; Science Press: Beijing, China; Missouri Botanical Garden Press: St. Louis, CA, USA, 2010; pp. 51–97.
2. Lee, H.-S. Insecticidal toxicities and essential oil compositions of *Zanthoxylum piperitum* and *Zanthoxylum schinifolium* fruits in Korea. *J. Essent. Oil-Bear. Plants* **2016**, *19*, 2065–2071. [CrossRef]
3. Phuyal, N.; Jha, P.K.; Raturi, P.P.; Gurung, S.; Rajbhandary, S. Essential oil composition of *Zanthoxylum armatum* leaves as a function of growing conditions. *Int. J. Food Prop.* **2019**, *22*, 1873–1885. [CrossRef]
4. Bubpawan, P.; Boonphong, S.; Sriwattanawarunyoo, C.; Udeye, V. Characterization of the Essential oil and Fatty oil from Makhwaen Fruit (*Zanthoxylum rhetsa* (Roxb.) DC). *NU. Int. J. Sci.* **2015**, *12*, 1–10.
5. Sriwichai, T.; Sookwong, P.; Siddiqui, M.W.; Sommano, S.R. Aromatic profiling of *Zanthoxylum myriacanthum* (makwhaen) essential oils from dried fruits using different initial drying techniques. *Ind Crop Prod.* **2019**, *133*, 284–291. [CrossRef]
6. Sriwichai, T.; Suksathan, R.; Charoenlertthanakit, N.; Sommano, S.R. *Zanthoxylum* spp.: A new potential sources of essential oil for the perfumery and pharmaceutical industries in Thailand. *Med. Plants* **2019**, *11*, 26–45.
7. Diao, W.R.; Hu, Q.P.; Feng, S.S.; Li, W.Q.; Xu, J.G. Chemical composition and antibacterial activity of the essential oil from green huajiao (*Zanthoxylum schinifolium*) against selected foodborne pathogens. *J. Agric. Food Chem.* **2013**, *61*, 6044–6049. [CrossRef] [PubMed]
8. Yamazaki, E.; Inagaki, M.; Kurita, O.; Inoue, T. Antioxidant activity of Japanese pepper (*Zanthoxylum piperitum* DC.) fruit. *Food Chem.* **2007**, *100*, 171–177. [CrossRef]
9. Lima, L.M.; Perazzo, F.F.; Carvalho, J.C.T.; Bastos, J.K. Anti-inflammatory and analgesic activities of the ethanolic extracts from *Zanthoxylum riedelianum* (Rutaceae) leaves and stem bark. *J. Pharm. Pharmacol.* **2007**, *59*, 1151–1158. [CrossRef] [PubMed]
10. Supabphol, R.; Tangjitjareonkun, J. Chemical constituents and biological activities of *Zanthoxylum limonella* (Rutaceae): A Review. *Trop. J. Pharm. Res.* **2015**, *13*, 2119. [CrossRef]
11. Chauhan, R.S.; Nautiyal, M.C.; Cecotti, R.; Mella, M.; Tava, A. Variation in the essential oil composition of *Angelica archangelica* from three different altitudes in Western Himalaya, India. *Ind. Crops Prod.* **2016**, *94*, 401–404. [CrossRef]
12. Jan, S.; Mir, J.I.; Shafi, W.; Faktoo, S.Z.; Singh, D.B.; Wijaya, L.; Alyemeni, M.N.; Ahmad, P. Divergence in tissue-specific expression patterns of genes associated with the terpeniod biosynthesis in two oregano species *Origanum vulgare* L., and *Origanum majorana*. *Ind. Crops Prod.* **2018**, *123*, 546–555. [CrossRef]
13. Charoensup, R.; Duangyod, T.; Phuneerub, P.; Singharachai, C. Pharmacognostic specifcation of *Zanthoxylum limonella* (Dennst.) Alston: Fruits and seeds in Thailand. *J. Adv. Pharm. Technol. Res.* **2016**, *7*, 134–138. [CrossRef]
14. Tangpao, T.; Chung, H.H.; Sommano, S.R. Aromatic profiles of essential oils from five commonly used Thai basils. *Foods* **2018**, *7*, 175. [CrossRef]
15. Zhang, M.; Wang, J.; Zhu, L.; Li, T.; Jiang, W.; Zhou, J.; Peng, W.; Wu, C. *Zanthoxylum bungeanum* Maxim. (Rutaceae): A systematic review of its traditional uses, botany, phytochemistry, pharmacology, pharmacokinetics, and toxicology. *Int. J. Mol. Sci.* **2017**, *18*, 2172. [CrossRef] [PubMed]
16. Haq, Q.; Hussain, T.; Kumar, A. Molecular markers: A tool to identify hidden science with especial emphasis on agricultural crops. *Int. J. Biol. Res.* **2016**, *1*, 50–59.
17. Wang, X.-Y.; Zheng, S.-H.; Liu, Y.; Han, J.-P. ITS2, a better DNA barcode than its in identification of Species in *Artemisia* L. *Chin. Herb. Med.* **2016**, *8*, 352–358. [CrossRef]
18. Yuan, T.; Jiao, Y.; de Jong, S.; Ophoff, R.A.; Beck, S.; Teschendorff, A.E. An integrative multi-scale analysis of the dynamic DNA methylation landscape in aging. *PLoS Genet* **2015**, *11*, e1004996. [CrossRef] [PubMed]
19. Roy, S.; Tyagi, A.; Shukla, V.; Kumar, A.; Singh, U.M.; Chaudhary, L.B.; Datt, B.; Bag, S.K.; Singh, P.K.; Nair, N.K.; et al. Universal plant DNA barcode loci may not work in complex groups: A case study with Indian berberis species. *PLoS ONE* **2010**, *5*, e13674. [CrossRef] [PubMed]
20. Muellner, A.N.; Schaefer, H.; Lahaye, R. Evaluation of candidate DNA barcoding loci for economically important timber species of the mahogany family (Meliaceae). *Mol. Ecol. Resour.* **2011**, *11*, 450–460. [CrossRef]
21. Singh, H.K.; Parveen, I.; Raghuvanshi, S.; Babbar, S.B. The loci recommended as universal barcodes for plants on the basis of floristic studies may not work with congeneric species as exemplified by DNA barcoding of Dendrobium species. *BMC Res. Notes* **2012**, *5*, 42. [CrossRef]
22. Suksathan, R.; Trisonthi, C.; Trisonthi, P.; Wangpakapattanawong, P. Notes on spice plants in the genus *Zanthoxylum* (Rutaceae) in Northern Thailand. *Thai. For. Bull. Bot.* **2009**, *37*, 197–204.

23. Rana, V.S.; Blazquez, M.A. Volatile constituents of the seed coat of *Zanthoxylum rhetsa* (Roxb.) DC. *J. Essent. Oil Res.* **2010**, *22*, 430–432. [CrossRef]
24. Das Gupta, D.; Mandi, S.S. Species Specific AFLP Markers for authentication of *Zanthoxylum acanthopodium* and *Zanthoxylum oxyphyllum*. *J. Med. Plants Stud.* **2013**, *1*, 1–9.
25. Feng, S.; Liu, Z.; Chen, L.; Hou, N.; Yang, T.; Wei, A. Phylogenetic relationships among cultivated Zanthoxylum species in China based on cpDNA markers. *Tree Genet. Genomes* **2016**, *12*, 45. [CrossRef]
26. Medhi, K.; Sarmah, D.K.; Deka, M.; Bhau, B.S. High gene flow and genetic diversity in three economically important *Zanthoxylum* spp. of upper Brahmaputra valley zone of NE India using molecular markers. *Meta Gene* **2014**, *2*, 706–721. [CrossRef] [PubMed]
27. Tatsadjieu, L.N.; Essia Ngang, J.J.; Ngassoum, M.B.; Etoa, F.X. Antibacterial and antifungal activity of *Xylopia aethiopica*, *Monodora myristica*, *Zanthoxylum xanthoxyloides* and *Zanthoxylum leprieurii* from Cameroon. *Fitoterapia* **2003**, *74*, 469–472. [CrossRef]
28. Jullian, V.; Bourdy, G.; Georges, S.; Maurel, S.; Sauvain, M. Validation of use of a traditional antimalarial remedy from French Guiana, *Zanthoxylum rhoifolium* Lam. *J. Ethnopharmacol.* **2006**, *106*, 348–352. [CrossRef] [PubMed]
29. Chambers, E.; Sanchez, K.; Phan, U.X.T.; Miller, R.; Civille, G.V.; Di Donfrancesco, B. Development of a "living′ lexicon for descriptive sensory analysis of brewed coffee. *J. Sens. Stud.* **2016**, *31*, 465–480. [CrossRef]
30. Taraj, K.; Delibashi, A.; Andoni, A.; Lazo, P.; Kokalari (Teli), E.; Lame, A.; Xhaxhiu, K.; Çomo, A. Extraction of chamomile essential oil by subcritical CO2 and its analysis by UV-VIS spectrophotometer. *Asian J. Chem.* **2013**, *25*, 7361–7364. [CrossRef]
31. Dodoš, T.; Rajčević, N.; Janaćković, P.; Vujisić, L.; Marin, P.D. Essential oil profile in relation to geographic origin and plant organ of *Satureja kitaibelii* Wierzb. ex Heuff. *Ind. Crops Prod.* **2019**, *139*, 111549. [CrossRef]
32. Costa, L.F.P.B.; Taleb Contini, S.H.; Teixeira, S.P.; Freitas, M.d.F.; França, S.d.C.; Bertoni, B.V.; Pereira, A.M.S. Secretory structures and chemical composition of the essential oil from leaves of *Myrsine leuconeura* Mart. (Primulaceae). *Flora* **2020**, *262*, 151496. [CrossRef]
33. Dhami, A.; Palariya, D.; Singh, A.; Kumar, R.; Prakash, O.; Kumar, R.; Pant, A. Chemical composition, antioxidant, in vitro anti-inflammatory and antibacterial activity of seeds essential oil of *Zanthoxylum armatum* DC. Collected from two different altitudes of Kumaun region, Uttarakhand. *Int. J. Chem. Stud.* **2018**, *6*, 363–370.
34. Phuyal, N.; Jha, P.K.; Prasad Raturi, P.; Rajbhandary, S. *Zanthoxylum armatum* DC.: Current knowledge, gaps and opportunities in Nepal. *J. Ethnopharmacol.* **2019**, *229*, 326–341. [CrossRef]
35. Shalabh, G.; Gautam, B.; Andola, H.C. Altitude dependent variation in essential oil content in bark of dioecious plants of *Zanthoxylum alatum* Roxb. *J. Indian Bot. Soc.* **2011**, *90*, 175–177.
36. Freitas, J.V.B.; Alves Filho, E.G.; Silva, L.M.A.; Zocolo, G.J.; de Brito, E.S.; Gramosa, N.V. Chemometric analysis of NMR and GC datasets for chemotype characterization of essential oils from different species of *Ocimum*. *Talanta* **2018**, *180*, 329–336. [CrossRef] [PubMed]
37. Xiang, L.; Liu, Y.; Xie, C.; Li, X.; Yu, Y.; Ye, M.; Chen, S. The Chemical and Genetic Characteristics of Szechuan Pepper (*Zanthoxylum bungeanum* and *Z. armatum*) Cultivars and Their Suitable Habitat. *Front. Plant. Sci.* **2016**, *7*, 467. [CrossRef] [PubMed]
38. Ji, Y.; Li, S.; Ho, C.-T. Chemical composition, sensory properties and application of Sichuan pepper (*Zanthoxylum* genus). *Food Sci. Hum. Wellness* **2019**, *8*, 115–125. [CrossRef]
39. Zhang, L.-L.; Zhao, L.; Wang, H.-Y.; Zhi, R.-C.; Shi, B.-L.; Xie, N. Determination of Recognition Threshold and Just Noticeable Difference in the Sensory Perception of Pungency of *Zanthoxylum bangeanum*. *Int. J. Food Prop.* **2016**, *19*, 1044–1052. [CrossRef]
40. Maisnam, D.; Rasane, P.; Dey, A.; Kaur, S.; Sarma, C. Recent advances in conventional drying of foods. *J. Food Process. Preserv.* **2017**, *1*, 25–34.
41. Samfira, I.; Rodini, S.; Petrache, P.; Cristina, R.T.; Butu, M.; Butnariu, M. Characterization and identity confirmation of essential oils by mid infrared absorption spectrophotometry. *Dig. J. Nanomater. Biostruct.* **2015**, *10*, 557–566.
42. Sommano, S.; Saratan, N.; Suksathan, R.; Pusadee, T. Chemical composition and comparison of genetic variation of commonly available Thai garlic used as food supplement. *J. Appl Bot. Food Qual.* **2016**, *89*, 235–242.
43. Gao, Y.; Zhou, P.; Mao, L.; Zhi, Y.-E.; Shi, W.-J. Assessment of effects of heavy metals combined pollution on soil enzyme activities and microbial community structure: Modified ecological dose-response model and PCR-RAPD. *Environ. Earth Sci.* **2009**, *60*, 603–612. [CrossRef]
44. Tutunoglu, B.; Aksoy, Ö.; Ozbek, R.; Uckan, F. The effects of gibberellic acid on *Allium cepa* root tip meristematic cells. *Biol. Plant.* **2019**, *63*, 365–370. [CrossRef]
45. Quang, H.; Nhan, B.; Đức, T.; Giang, T.; Pham, T. Study on genetic diversity of *Curculigo orchioides* Gaertn populations from Vietnam, an endangered medicinal herb. *Plant Cell Biotechnol. Mol. Biol.* **2018**, *19*, 293–301.
46. Das, S.; Singh, Y.; Negi, Y.; Srivastava, P. Genetic variability in different growth forms of *Dendrocalamus strictus*: Deogun revisited. *N. Z. J. For. Sci.* **2017**, *47*, 23. [CrossRef]

MDPI
St. Alban-Anlage 66
4052 Basel
Switzerland
Tel. +41 61 683 77 34
Fax +41 61 302 89 18
www.mdpi.com

Plants Editorial Office
E-mail: plants@mdpi.com
www.mdpi.com/journal/plants

www.ingramcontent.com/pod-product-compliance
Lightning Source LLC
Chambersburg PA
CBHW041955260926
51850CB00019B/2894
* 9 7 8 3 0 3 6 5 6 9 8 2 6 *